THE NOISY OSCILLATOR

The First Hundred Years, From Einstein Until Now

THE NOISY OSCILLATOR

The First Hundred Years,
From Einstein Until Now

Moshe Gitterman

Bar-Ilan University, Israel

 World Scientific

NEW JERSEY • LONDON • SINGAPORE • BEIJING • SHANGHAI • HONG KONG • TAIPEI • CHENNAI

Published by

World Scientific Publishing Co. Pte. Ltd.

5 Toh Tuck Link, Singapore 596224

USA office: 27 Warren Street, Suite 401-402, Hackensack, NJ 07601

UK office: 57 Shelton Street, Covent Garden, London WC2H 9HE

British Library Cataloguing-in-Publication Data
A catalogue record for this book is available from the British Library.

THE NOISY OSCILLATOR
The First Hundred Years, From Einstein Until Now

ISBN-13 978-981-256-512-9
ISBN-10 981-256-512-4

Printed in Singapore

In memory of my beloved Neta

Preface

At one time I worked on the first heart sound produced by the left ventricular. The existence of a high frequency component was one of the predictions of the new theory, and I was very interested to get an experimental verification of this prediction. It was a reason for my talk for the group of cardiologists in one of the hospitals. I started my talk by saying: "Let us consider the left ventricular as a sphere or cylinder...". At that moment I lost my audience who knew (as I did) that the left ventricular has neither the form of a sphere nor that of a cylinder. On the other hand, I knew how to build the model for the simplest geometry of sphere or cylinder. When an average physicists comes up against a new problem, he immediately starts to apply to this problem all the methods and ideas he knew from his previous experience[1].

The first step is the choice of some simplified model (geometric form of the left ventricular in my case). The simplest, the most general and the most used model is that of a harmonic oscillator. This model is used for the description of different phenomena in mechanics, optics, acoustics, electronics, engineering, etc. [1]. In fact, it has been applied everywhere, from quarks to cosmology. Moreover, a person who is worried by oscillations of prices in the stock market can be relaxed by classical music produced by the oscillations of string instruments.

---

[1]Of course, the great physicists do not come under these headings - recall A. Einstein with general relativity, P. Dirac with relativistic quantum mechanics, or L. Onsager with the two-dimensional Ising problem. In fact, they created not only the new physics, but also the new mathematics

Wandering about the Internet one can find many curious facts. Although the ancient Greeks already had a general idea of oscillations and used them in musical instruments, the first practical application of an oscillating system took place in 1602 by a physician in Venice named Santorio who had heard from his great friend Galileo about the general laws of the oscillations of a pendulum. Mr. Santorio called this system "pulsilogium" and used it to measure the pulse of his patients. Many other applications have been found in the last 400 years...

Two main features characterize this book. Firstly, the book contains a comprehensive description of all "oscillator-like" stochastic differential equations which were studied until 2004, and it can serve as a starting point for researchers and engineers who meet these equations in their work. The second characteristic feature of this book is its simplicity and small volume.

Dozens of existing books and many hundreds of articles create a serious problem that confronts the author of a new book on stochastic differential equation. The problem consists of a difficult decision about restricting the consideration to some specific problem. My decision was to concentrate on a single stochastic one-dimensional classical oscillator, omitting thereby the quantum problem, interactions of oscillators and higher dimensions as well as the fascinating problem of deterministic chaos.

I decided not only to avoid all rigorous mathematical proofs and statements, but also to drop the large body of applications and the traditional comprehensive introduction to the mathematical theory of random processes. All necessary explanations are given in the appropriate sections of the book. If the material presented demands complicated calculations, I covered only the qualitative results referring for details to the original articles. Only the general knowledge of mathematical physics is required of the reader.

This book is devoted to "noisy" equations, i.e., equations which contain random forces, although for the reader's convenience, chapter 1 contains a short review of the deterministic equations and the types of the second order (underdamped) differential equations involving additive and/or multiplicative noise which are considered, along with their simplified versions (first order differential equation) in great detail in this book. Chapter 2, which is devoted to the short general descriptions of noise, is required for understanding the ensuing material. In chapter 3 we consider the Brownian motion, thereby paying respect to Einstein [2], Smoluchowski [3] and Langevin [4] who were the first to introduce a random force in deterministic equations for velocity and for coordinate. The following chapters contain detailed analyses of the overdamped (chapters 4 − 7) and underdamped (chapters 8 − 12) noisy oscillators.

Contents

xi

Chapter 1

Deterministic and random oscillators

This chapter reviews all types of differential equations which will be analyzed later on in more detail. We present here the second order (underdamped) differential equations. The first order (overdamped) equations, which we will consider first can be obtained from it by leaving only the first derivative in these equations.

1.1 Simple harmonic oscillator

Such an oscillator is a simple system where the net force is directly proportional to the displacement x of the mass m from the equilibrium position $x = 0$ and pointing in the opposite direction (the Hook's law). According to Newton's law of motion,

$$m\frac{d^2x}{dt^2} = -kx \quad \text{or} \quad m\frac{d^2x}{dt^2} + kx = 0. \tag{1.1}$$

The solution of this equation has a form

$$x = C\cos(\omega_0 t + \phi), \tag{1.2}$$

where the angular frequency $\omega_0 = \sqrt{k/m}$ depends only upon the parameters of the system, while the amplitude A and the phase ϕ are constants determined by the initial displacement $x(t = 0)$ and velocity $\frac{dx}{dt}(t = 0)$.

1

1.2   Damped harmonic oscillator

The preceding analysis can be supplemented by a dissipative force which points in the opposite direction and is usually assumed to be proportional to the velocity,

$$m\frac{d^2x}{dt^2} = -kx - 2\gamma\frac{dx}{dt} \quad \text{or} \quad m\frac{d^2x}{dt^2} + 2\gamma\frac{dx}{dt} + kx = 0. \quad (1.3)$$

The solution of equation (1.3) for $k > \gamma$ has the form

$$x = C\exp\left(-\frac{\gamma t}{m}\right)\cos\left(\omega_1 t + \phi\right), \quad (1.4)$$

with $\omega_1 = \sqrt{\omega_0^2 - \left(\frac{\gamma}{m}\right)^2}$.

1.3   Driven harmonic oscillator

Another generalization of the equations for a simple harmonic oscillator (1.1) and a damped harmonic oscillator (1.3) is one driven by some time-dependent external force, the simplest form of which is a sinusoidal function of time, i.e.,

$$m\frac{d^2x}{dt^2} + kx = A\cos\left(\Omega t + \beta\right). \quad (1.5)$$

This has a driven solution of the form

$$x = \frac{A}{m\left(\omega_0^2 - \Omega^2\right)}\cos\left(\Omega t + \beta\right). \quad (1.6)$$

When the external frequency Ω approaches the intrinsic frequency ω_0, the steady state amplitude approaches infinity (dynamic resonance). Classic demonstrations of this dynamic resonance are two architectural flaws in the US. The first is the Takoma bridge which was destroyed by the wind's force with the resonance frequency, and the second is the Paramount Communication building in New York which was transformed into a luxury apartment building. Due to the winds the top floors were twisted and windows pried loose from their casements.

1.4 Driven, damped harmonic oscillator

On combining the damping and driving forces, one arrives at the following equation

$$m\frac{d^2x}{dt^2} + kx + 2\gamma\frac{dx}{dt} = A\cos\left(\Omega t + \beta\right). \tag{1.7}$$

The general solution of equation (1.7) is a sum of a transient and a steady state solutions

$$x = C\exp\left(-\frac{\gamma t}{m}\right)\cos\left(\omega_1 t + \phi\right)$$

$$+ \frac{A}{m\sqrt{\left(\omega_0^2 - \Omega^2\right)^2 + \frac{4\gamma^2\Omega^2}{m^2}}}\sin\left(\Omega t + \phi + \beta\right) \tag{1.8}$$

with $\beta = \tan^{-1}\frac{\omega_0^2 - \Omega^2}{2\gamma\Omega}$.

1.5 Nonlinear oscillator in a double-well potential

Up to now, we have considered the linear oscillator with the restoring force f proportional to the displacement x, $f \approx -kx$. In a more general case this dependence can be nonlinear, $f \approx ax - bx^n$. Hereafter, we restrict our attention to the most interesting cases $n = 2$ and $n = 3$.

The linear force in equations (1.1), (1.3), (1.7) corresponds to the potential $U = kx^2/2$. For the more complicated case of a double-well potential energy $U = -ax^2/2 + bx^4/4$, with two minima located at $x = \pm\sqrt{a/b}$ and a maximum at $x = 0$, the equation of motion has the form

$$m\frac{d^2x}{dt^2} - ax + bx^3 = 0. \tag{1.9}$$

1.6   Nonlinear oscillator in a single-well potential

Another much used form of the potential energy is one of the form $U = -ax^2/2 + bx^3/3$, with an unstable minimum at $x = a/b$, which results in the following equation of motion

$$m\frac{d^2x}{dt^2} - ax + bx^2 = 0. \tag{1.10}$$

Both equations (1.9) and (1.10) can be easily generalized to include a damping term and an external force. However, one cannot find the exact solutions of these equations even for a simplified form of these equations.

1.7   Harmonic oscillator with external noise

In the foregoing we considered a pure mechanical system (zero temperature). However, for all finite temperatures the dynamic equations (1.1), (1.4) have to be supplemented by thermal noise, i.e.,

$$m\frac{d^2x}{dt^2} + 2\gamma\frac{dx}{dt} + kx = \xi(t), \tag{1.11}$$

where $\xi(t)$ is the random variable with zero mean $\langle \xi(t) \rangle = 0$ and the variance $\langle \xi^2(t) \rangle$ which for the thermal noise must satisfy the fluctuation-dissipation theorem [5], $\langle \xi^2(t) \rangle = 4\gamma\kappa T$, where κ is the Boltzmann constant. The noise acting on a system can also be external (and not thermal noise) with no special requirement for the value of $\langle \xi^2(t) \rangle$. Still another way to justify the validity of equation (1.11) is as follows: in considering only one (slow) mode $x(t)$ of a complex system, one may take into account the influence of other (fast) modes by introducing a random force into the dynamic equations.

1.8   Brownian motion

Equation (1.11) with $m = 0$ and x meaning the particle's velocity v describes Brownian motion, where the force acting on the Brownian particle consists of the systematic force $-kv$ and the random force $\xi(t)$

$$\gamma\frac{dv}{dt} + kv = \xi(t).\tag{1.12}$$

Note, that caution is required in going from the oscillator equation to that of the Brownian particle, since the simple substitution of $m \to 0$ in equation (1.11) will decrease the order of the differential equation, and one has to use the singular perturbation theory [6].

1.9 Harmonic oscillator with random frequency

The random force $\xi(t)$ enters equation (1.11) additively. When the noise has an external origin rather than an internal one, for instance if it arises from the fluctuations of the potential energy $U = \frac{kx^2[1+\xi(t)]}{2}$, the equation of motion of an oscillator with an external periodic force $A\sin(\Omega t)$, will take the following form:

$$m\frac{d^2x}{dt^2} + 2\gamma\frac{dx}{dt} + k\left[1 + \xi(t)\right]x = A\sin(\Omega t).\tag{1.13}$$

The internal dynamics of a force-free harmonic oscillator ($A = 0$) with random frequency is a subject that has been extensively investigated in different fields, including physics (on-off intermittence [7], dye lasers [8], wave propagation in random media [9], turbulent flows [10]), biology (population dynamics [11]), economics (stock market prices [12]) and so on.

Note that for $A = \gamma = 0$ and for x and t being replaced by the electric field and coordinate, respectively, equation (1.13) transforms into the Maxwell equation with a random dielectric constant which has also been intensively studied [13]. Recently we considered equation (1.13) with an external force [14], [15].

1.10   Harmonic oscillator with random damping

Another possibility for the generalization of the dynamic equation (1.7) is the incorporation of random damping

$$m\frac{d^2x}{dt^2} + 2\gamma\left[1 + \xi\left(t\right)\right]\frac{dx}{dt} + kx = A\sin\left(\Omega t\right) \qquad (1.14)$$

The first time this equation (with $A = 0$) was used [16] for the problem of water waves influenced by a turbulent wind field. However, this equation (with the coordinate x and time t replaced by the order parameter and the coordinate, respectively) is transformed into the Ginzburg-Landau equation with a convective term which describes phase transitions in moving systems [17]. There is an increasing number of problems where the particles advected by the mean flow pass through the region under study. These include problems of phase transition under shear [18], open flows of liquids [19], Rayleigh-Benard and Taylor-Couette problems in fluid dynamics [20], dendritic growth [21], chemical waves [22], and the motion of vortices [23].

Notice the important difference between the linear equation (1.11) with an additive noise and equations (1.13), (1.14) with multiplicative noise. The latter show, in fact, "quasi-nonlinear behavior" including the stochastic resonance phenomenon [14], [24].

Chapter 2

White and colored noise

In the following we will consider noise $\xi(t)$ with $\langle \xi(t) \rangle = 0$ and the correlator

$$\langle \xi(t_1)\xi(t_2) \rangle = r \mid t_1 - t_2 \mid \equiv r(z). \tag{2.1}$$

Two integrals of (2.1) characterize the fluctuations: the strength of the noise D

$$D = \frac{1}{2} \int_0^\infty \langle \xi(t)\xi(t+z) \rangle \, dz, \tag{2.2}$$

and the correlation time τ,

$$\tau = \frac{1}{D} \int_0^\infty z \langle \xi(t)\xi(t+z) \rangle \, dz. \tag{2.3}$$

Traditionally one considers two different forms of noise, white and colored noise. For white noise the function $r \mid t_1 - t_2 \mid$ has the form of a delta-function,

$$\langle \xi(t_1)\xi(t_2) \rangle = 2D\delta(t - t_1). \tag{2.4}$$

The name "white" noise comes from the fact that the Fourier transform of (2.4) is "white" being constant without any characteristic frequency. Equation (2.4) means that noises $\xi(t_1)$ and $\xi(t_2)$ are statistically independent, no matter how close t_1 and t_2 are. This extreme assumption, which leads to a non-physical infinite value of $\langle \xi^2(t) \rangle$ in (2.4), means, in fact, that the correlation time τ is not zero as assumed in (2.4), but smaller than all the other characteristic times in a problem. We will return to this subject in the next chapter on examining the Fokker-Planck equation for multiplicative noise.

All other non-white noises are called colored noise. A widely used noise is the Ornstein-Uhlenbeck exponentially correlated noise which can be written in two forms,

$$\langle \xi(t)\,\xi(t_1) \rangle = \sigma^2 \exp\left[-\lambda\,|\,t - t_1\,|\right], \qquad (2.5)$$

or

$$\langle \xi(t)\,\xi(t_1) \rangle = \frac{D}{\tau} \exp\left[-\frac{|\,t - t_1\,|}{\tau}\right]. \qquad (2.6)$$

White noise (2.4) is defined by its strength D while the Ornstein-Uhlenbeck noise is characterized by two parameters, λ and σ^2, or τ and D. The transition from the Ornstein-Uhlenbeck noise to white noise (2.4) is performed by the limit $\tau \to 0$ in (2.6), or by $\sigma^2 \to \infty$ and $\lambda \to \infty$ in (2.5) in such a way that $\sigma^2/\lambda = 2D$.

A slightly generalized form of the Ornstein-Uhlenbeck noise is the so-called narrow-band colored noise with a correlator of the form,

$$\langle \xi(t)\,\xi(t_1) \rangle = \sigma^2 \exp\left[-\lambda\,|\,t - t_1\,|\right] \cos\left(\omega_2\,|\,t - t_1\,|\right). \qquad (2.7)$$

There are different forms of colored noises which we will briefly review in this chapter.

2.1   Dichotomous noise

A special type of colored noise with which we shall be concerned is the symmetric dichotomous noise (random telegraph signal) where the random variable $\xi(t)$ may take one of the values $\xi = \pm\sigma$ with the mean waiting time $(\lambda/2)^{-1}$ in each of these two states. Like the Ornstein-Uhlenbeck noise, dichotomous noise is characterized by the correlators (2.5)-(2.6).

In what follows we will use the Furutzu-Novikov-Shapiro-Loginov procedure [25], [26] for splitting the higher order correlations, which for exponentially correlated noise yields

$$\frac{d}{dt} \langle \xi \cdot g \rangle = \left\langle \xi \frac{dg}{dt} \right\rangle - \lambda \langle \xi \cdot g \rangle,\tag{2.8}$$

where g is some function of noise, $g = g\{\xi\}$. If $\frac{dg}{dt} = B\xi$, then Eq. (2.8) becomes

$$\frac{d}{dt} \langle \xi \cdot g \rangle = B \langle \xi^2 \rangle - \lambda \langle \xi \cdot g \rangle,\tag{2.9}$$

and for white noise ($\xi^2 \to \infty$ and $\lambda \to \infty$ with $\xi^2/\lambda = 2D$), one gets

$$\langle \xi \cdot g \rangle = 2BD.\tag{2.10}$$

2.2 White Poisson noise

We define the white Poisson noise as a train of δ-functions at random times t_i, i.e.

$$\xi(t) = \sum_{i=1}^{n(t)} \omega_i \delta(t - t_i) - \lambda \omega \tag{2.11}$$

where $n(t)$ is the Poisson process with parameter λ which controls the average time difference between two pulses. The amplitude of the pulses ω_i is exponentially distributed with the mean value ω,

$$\rho(\omega_i) = \frac{\exp\left(-\frac{\omega_i}{\omega}\right)}{\omega}.\tag{2.12}$$

Here λ is the mean density of t_i, and the last term in (2.11) is added to ensure a zero value for the mean noise, $\langle \xi(t) \rangle = 0$.

2.3 Shot noise

The generalization of (2.11) results in the so-called shot noise,

$$\xi(t) = \sum_i \omega_i f(t - t_i) - g(t) \tag{2.13}$$

where ω_i and t_i are random variables described by the Poisson process (2.12), where $f(t)$ is some response function, and $g(t)$ is the so-called deterministic compensator responsible for the condition $\langle \xi(t) \rangle = 0$.

If the "width" $t - t_i$ of $f(t - t_i)$ is small compared with other characteristic times of a problem, one can simplify equation (2.13) by replacing $f(t - t_i)$ by the δ-function (white shot noise),

$$\xi(t) = \sum_i \omega_i \delta(t - t_i) - g(t). \qquad (2.14)$$

Chapter 3

Brownian motion

3.1 In the beginning...

This name came from the observation of the peculiar erratic motion of a mesoscopic particle of a micron size floating among microscopic particles performed by Robert Brown in 1827 [27]. The theoretical description of this phenomenon was given by Einstein [2], Smoluchowski [3], and Langevin [4] who introduced the molecular-kinetics corrections to deterministic equations. Langevin considered the equation of motion for the position of the Brownian particle while Einstein and Smoluchowski that for its velocity[1]. They assume that the total force acting on the Brownian particle is decomposed into a systematic force (viscous friction proportional to velocity, $f = -\gamma v$), and a fluctuation force $\xi(t)$ exerted on the Brownian particle by the molecules of the surrounding medium. This fluctuation force comes from the different number of molecular collisions with a Brownian particle from two opposite sites, resulting in its random motion. Then, for a one-dimensional system Newton's law gives

$$m\frac{dv}{dt} = -\gamma v + \xi(t).\qquad(3.1)$$

The simplest assumption of white noise (2.4) has been made in the original articles. The more general case will be considered in the next section.

---

[1]For the velocity they used the twenty four-letter German word "Bewegungs-geschwindinkeit"!

Both terms in the right-hand side of equation (3.1) are connected with collisions between a Brownian particle and molecules of the surrounding media. The separation of this interaction into two parts is somewhat artificial, and it is not surprising, therefore, that their characteristics are connected by the fluctuation-dissipation theorem [5].

Equation (3.1) has a simple solution

$$v = v\,(t=0)\exp\left(-\frac{\gamma t}{m}\right) + \frac{1}{m}\int_0^t \exp\left[-\frac{\gamma}{m}(t-t_1)\right]\xi\,(t_1)\,dt_1. \quad (3.2)$$

Integration of (3.2) over t gives

$$x = x\,(t=0) + \frac{mv\,(t=0)}{\gamma}\left[1 - \exp\left(-\frac{\gamma t}{m}\right)\right]$$

$$+ \frac{1}{\gamma}\int_0^t \left\{1 - \exp\left[-\frac{\gamma}{m}(t-t_1)\right]\right\}\xi\,(t_1)\,dt_1. \quad (3.3)$$

Using (3.3) one obtains for the mean-square displacement

$$\langle\sigma^2\rangle \equiv \left\langle[x\,(t) - x\,(t=0)]^2\right\rangle = \frac{m^2 v\,(t=0)^2}{\gamma^2}\left[1 - \exp\left(-\frac{\gamma}{m}t\right)\right]^2$$

$$+ \frac{1}{\gamma^2}\int_0^t dt_1 \int_0^t \left\{1 - \exp\left[-\frac{\gamma}{m}(t-t_1)\right]\right\}$$

$$\times \left\{1 - \exp\left[-\frac{\gamma}{m}(t-t_2)\right]\right\}\langle\xi\,(t_1)\,\xi\,(t_2)\rangle\,dt_2. \quad (3.4)$$

Substituting in (3.4) the correlator (2.4) of white noise, one gets

$$\langle\sigma^2\rangle = \frac{2Dt}{\gamma^2} + \frac{m^2 v\,(t=0)^2}{\gamma^2}\left[1 - \exp\left(-\frac{\gamma}{m}t\right)\right]^2$$

$$+ \frac{mD}{\gamma^3}\left[-3 + 4\exp\left(-\frac{\gamma}{m}t\right) - \exp\left(-\frac{2\gamma}{m}t\right)\right], \quad (3.5)$$

and for long times, $t \to \infty$,

$$\left\langle(\Delta x)^2\right\rangle = \frac{2Dt}{\gamma^2} = \frac{2\varkappa T}{\gamma}t. \quad (3.6)$$

The last equality in (3.6) follows from the fluctuation-dissipation theorem [5], $\langle \xi^2(t) \rangle = 2\gamma\kappa T$, where κ is the Boltzmann constant. The latter simply means that the power entering the system from the external random force must be entirely dissipated and given off to the thermostat in order that the equilibrium state of the system should not be disturbed. Analogously, one gets from (3.2) for the long time limit value of the mean velocity

$$\langle v^2 \rangle = \frac{\kappa T}{m}, \tag{3.7}$$

the well-known statistical mechanics equipartition theorem. The equilibrium distribution comes about as balance of two contrary processes in equation (3.1), the random force which tends to increase the velocity of the Brownian particle, and the damped force which tries to stop the particle.

3.2 Fokker-Planck equation

3.2.1 *Additive white noise*

The stochastic equation (3.1) which is called Langevin equation, describes the motion of an individual Brownian particle. The random force $\xi(t)$ in this equation causes the solution of this equation $v(t)$ to be random as well. Equivalently, one can consider an ensemble of Brownian particles and ask how many particles of this ensemble have at time t the velocities in the interval $(v, v + dv)$ described by the probability density $P(v,t)$. The deterministic equation for $P(v,t)$ is called the Fokker-Planck equation, and for the white noise (2.4) has the following form

$$\frac{\partial P(v,t)}{\partial t} = \frac{\partial}{\partial v}[\gamma v P] + D\frac{\partial^2 P}{\partial v^2}. \tag{3.8}$$

Multiplying (3.8) by v and v^2 and performing the integrations by parts on the right hand side of this equation, one gets for the first two moments $\langle v \rangle$, $\langle v^2 \rangle$

$$\langle v \rangle = v\,(0) \exp\left(-\frac{\gamma}{m}t\right),$$

$$\langle v^2 \rangle = v\,(0)^2 \exp\left(-\frac{2\gamma}{m}t\right) + \frac{D}{m\gamma}\left[1 - \exp\left(-\frac{2\gamma}{m}t\right)\right].$$

(3.9)

Let us demonstrate [28] the equivalence of the Fokker-Planck equation

$$\frac{\partial P\,(x,t)}{\partial t} = -\frac{\partial}{\partial x}\left[f\,(x)\,P\right] + D\frac{\partial^2 P}{\partial x^2}$$

(3.10)

and the Langevin equation

$$\frac{dx}{dt} = f\,(x) + \xi\,(t)$$

(3.11)

with an arbitrary (nonlinear) function $f\,(x)$ and white noise $\xi\,(t)$.

Integrating equation (3.11) between t and $t + \Delta t$ one gets

$$x\,(t+dt) - x\,(t) \equiv \Delta x = \int_t^{t+\Delta t} f\,[x\,(u)]\,du + \int_t^{t+\Delta t} \xi\,(u)\,du.$$

(3.12)

Averaging this equation over $\xi\,(t)$ at constant $x\,(t)$ results in

$$\langle \Delta x \rangle = f\,[x\,(t)]\,\Delta t + O\left[(\Delta t)^2\right]$$

(3.13)

which gives the first term in (3.10). Then,

$$\left\langle (\Delta x)^2 \right\rangle = \left\langle \left[\int_t^{t+\Delta t} f\,[x\,(u)]\,du\right]^2 \right\rangle$$

$$+ 2\int_t^{t+\Delta t} du \int_t^{t+\Delta t} dz \,\langle f\,[x\,(u)]\,\xi\,(z)\rangle$$

$$+ \int_t^{t+\Delta t} du \int_t^{t+\Delta t} dz \,\langle \xi\,(u)\,\xi\,(z)\rangle.$$

(3.14)

The first two terms in (3.14) are of the order of $(\Delta t)^2$ and can be neglected, just as in (3.13) while the last term gives $D\Delta t$ which shows the equivalence of the Langevin equation (3.11) and the Fokker-Planck equation (3.10). One additional comment is required to show the full equivalence of these equations. The Fokker-Planck equation defines the distribution function $P(x)$ which allows one to calculate all moments of $x(t)$ while the Langevin equation contains only the first two moments. To eliminate this contradiction we will consider only those delta-correlated noises (the so-called Gaussian white noises) for which the first two moments define all higher moments, namely all odd moments vanish, and all even moments are

$$\langle \xi(t_1) \xi(t_2) ... \xi(t_n) \rangle = \sum (2D)^n \delta(t_1 - t_2) ... \delta(t_{n-1} - t_n). \quad (3.15)$$

The summation in (3.15) extends over all different ways in which the indices $1, 2, ... n$ can be subdivided in pairs.

3.2.2 *Multiplicative white noise*

The random force entered equation (3.1) additively. The situation is more problematic for noise which appears multiplicatively in an equation of motion,

$$\frac{dx}{dt} = f[x(t)] + g[x(t)] \xi(t). \quad (3.16)$$

Usually the functions $f(x)$ and $g(x)$ are powers of x, $f \sim x^n$ and $g \sim x^m$. Then equation (3.16) takes the form

$$\frac{dx}{dt} = ax^n + bx^m \xi(t). \quad (3.17)$$

By introducing the new variable $y = x^{1-m}$ one can convert equation (3.17) with multiplicative noise to the form

$$\frac{dy}{dt} = a(1-m) y^{\frac{n-m}{1-m}} + b(1-m) \xi(t) \quad (3.18)$$

with additive noise.

As was discussed earlier after equation (2.4), the real "white noise" $\xi(t)$ could be expressible as a sequence of randomly arriving delta peaks of finite width which causes a jump in $x(t)$. It is not clear, therefore, at which moment one has to take the value of x in the function $g[x(t)]$ (at the beginning, in the middle or at the end of a jump). The latter decision will lead to the different forms of the Fokker-Planck equation. The three possibilities described above will lead to the Ito-Stratonovich [28] or kinetic [29] interpretation, respectively. Integrating equation (3.16) one gets, analogously to (3.12),

$$x\left(t+\Delta t\right)-x\left(t\right)$$
$$= f\left[x\left(t\right)\right]\Delta t + g\left[x\left(t\right)\right]\int_t^{t+\Delta t} \xi\left(u\right)du, \qquad\qquad \text{(Ito)}$$
$$x\left(t+\Delta t\right)-x\left(t\right)$$
$$= f\left[x\left(t\right)\right]\Delta t + g\left[\frac{x\left(t\right)+x\left(t+\Delta t\right)}{2}\right]\int_t^{t+\Delta t} \xi\left(u\right)du, \quad \text{(Stratonovich)}$$
$$x\left(t+\Delta t\right)-x\left(t\right)$$
$$= f\left[x\left(t\right)\right]\Delta t + g\left[x\left(t+\Delta t\right)\right]\int_t^{t+\Delta t} \xi\left(u\right)du, \qquad\qquad \text{(kinetic)}$$
$$\tag{3.19}$$

It follows from (3.19) [28] that the Fokker-Planck equation will be different in the Ito and Stratonovich interpretation

$$\frac{\partial P\left(v,t\right)}{\partial t} = -\frac{\partial}{\partial x}\left\{\left[f\left(x\right) + \nu Dg\left(x\right)\frac{dg}{dx}\right]P\right\} + D\frac{\partial^2}{\partial x^2}\left[g^2\left(x\right)P\right]$$
$$\tag{3.20}$$

with $\nu = 1$ (Stratonovich) and $\nu = 0$ (Ito).

Note that the Stratonovich interpretation is more natural since it follows from the usual calculus. Indeed, let us change the variables in equation (3.16)

$$x_1 = \int\frac{dx}{g\left(x\right)}; \quad f\left(x_1\right) = \frac{f\left(x\right)}{g\left(x\right)}; \quad P\left(x_1\right) = P\left(x\right)g\left(x\right) \qquad (3.21)$$

which gives

$$\frac{\partial P\left(x_1,t\right)}{\partial t} = -\frac{\partial}{\partial x_1}\left[f\left(x_1 P\right)\right] + D\frac{\partial^2 P}{dx_1^2} \qquad\qquad (3.22)$$

and, after transforming back to the original variable x, one gets the Stratonovich form of the Fokker-Planck equation (3.20) with $\nu = 1$.

Let us mention also an another way of getting the Fokker-Planck equation in the presence of both additive and multiplicative white noises [30]

$$\frac{dx}{dt} = f\left[x\left(t\right)\right] + g\left[x\left(t\right)\right]\xi\left(t\right) + \eta\left(t\right) \tag{3.23}$$

where

$$\langle\xi\left(t_1\right)\xi\left(t_2\right)\rangle = 2D\delta\left(t_1 - t_2\right); \quad \langle\eta\left(t_1\right)\eta\left(t_2\right)\rangle = 2\alpha\delta\left(t_1 - t_2\right). \tag{3.24}$$

Then, the Fokker-Planck equation (3.8) transforms into a stochastic equation

$$\frac{\partial P\left(x,t;\xi\right)}{\partial t} = -\frac{\partial}{\partial x}\left[\{f\left(x\right) + g\left[x\left(t\right)\right]\xi\left(t\right)\}P\right] + \alpha\frac{\partial^2 P}{\partial x^2}. \tag{3.25}$$

We are interested in the probability density $P\left(x,t\right)$, defined as the average of $P\left(x,t;\xi\right)$ over all realizations of $\xi\left(t\right)$. Averaging (3.25) one gets

$$\frac{\partial P\left(x,t\right)}{\partial t} = -\frac{\partial}{\partial x}\left[f\left(x\right)P\right] + \alpha\frac{\partial^2 P}{\partial x^2} - \frac{\partial}{\partial x}\left\{g\left[x\left(t\right)\right]\langle\xi\left(t\right)P\left(x,t;\xi\right)\rangle_\xi\right\}. \tag{3.26}$$

The last term in (3.26) can be computed using the Novikov-Furutzu theorem for a Gaussian process [25]

$$\langle\xi\left(t\right)P\left(x,t;\xi\right)\rangle_\xi = \int_0^t dz\,\langle\xi\left(t\right)\xi\left(z\right)\rangle\left\langle\frac{\partial P\left(x,t;\xi\right)}{\partial\xi\left(z\right)}\right\rangle$$

$$= -D\frac{\partial}{\partial x}\left\{g\left[x\left(t\right)\right]P\left(x,t\right)\right\}. \tag{3.27}$$

Inserting this result in (3.26) one gets

$$\frac{\partial P\left(x,t\right)}{\partial t} = \frac{\partial}{\partial x}\left\{\left[f\left(x\right) + \alpha\frac{\partial g}{\partial x}\right]P\left(x,t\right)\right\}$$

$$+ \frac{\partial^2}{\partial x^2}\left(\{\alpha + Dg^2[x\left(t\right)]\}P\left(x,t\right)\right) \tag{3.28}$$

which for $\alpha = 0$ coincides with (3.20).

3.2.3   *Colored noise: unified colored noise approximation (UCNA)*

There is no rigorous way to go from the Langevin equation

$$\frac{dx}{dt} = f(x) + \xi_1(t) g_1(x) + \xi_2(t) g_2(x) \tag{3.29}$$

with colored noises $\xi_{1,2}(t)$ to the appropriate Fokker-Planck equation. We consider the best known approximation, the UCNA [31] scheme for the Ornstein-Uhlenbeck noise with the correlator

$$\langle \xi_i(t_1) \xi_i(t_2) \rangle = \frac{D_i}{\tau_i} \exp\left(-\frac{|t_1 - t_2|}{\tau_i}\right), \quad i = 1, 2. \tag{3.30}$$

One can rewrite equations (3.29)-(3.30) in the following form [32]

$$\frac{dx}{dt} = f(x) + \xi_1(t) g_1(x) + \xi_2(t) g_2(x),$$
$$\frac{d\xi_i}{dt} = -\frac{1}{\tau_i}\xi_i + \frac{\sqrt{D_i}}{\tau_i}\varepsilon_i(t), \tag{3.31}$$

where $\varepsilon(t)$ is white noise, $\langle \varepsilon_i(t_1) \varepsilon_i(t_2) \rangle = 2\delta(t_1 - t_2)$. On eliminating ξ_i from equations (3.31) and introducing the new variable $s = \tau^{-\frac{1}{2}}t$, one gets

$$\frac{d^2 x}{ds^2} + \eta(x,\tau)\frac{dx}{ds} - f(x) = \frac{\sqrt{D}}{\tau^{1/4}}\varepsilon(s), \tag{3.32}$$

where for additive noise ($g_1 = 1$, $g_2 = 0$) the nonlinear damping $\eta(x,\tau)$ reads

$$\eta(x,\tau) = \frac{1}{\sqrt{\tau}} + \sqrt{\tau}\left(-\frac{df}{dx}\right), \tag{3.33}$$

and for multiplicative noise ($g_2 = 0$),

$$\eta(x,\tau) = \frac{1}{\sqrt{\tau}} + \sqrt{\tau}\left[-\frac{df}{dx} + f(x)\frac{d(\ln g_1(x))}{dx}\right]. \tag{3.34}$$

If the noise-free equation (3.29) is stable, then $\frac{df}{dx} < 0$, and the expressions in the square brackets in (3.33) and (3.34) are positive. Moreover, $\eta(x,\tau)$ approaches infinity for both $\tau \to 0$ and $\tau \to \infty$. Therefore, for large positive damping one can neglect the second derivatives in equation (3.32), and rewrite this equation in terms of the original variable $t = \sqrt{\tau} s$ as

$$\eta\left(x,t\right)\frac{dx}{dt} = f\left(x\right) + \left[g_1\left(x\right)D_1 + g_2\left(x\right)D_2\right]\varepsilon\left(t\right) \tag{3.35}$$

with

$$\eta\left(x,t\right) = 1 - \sum_{i=1,2}\tau_i\left[\frac{df}{dx} - \frac{d\ln g_i}{dx_i}f\left(x\right)\right]. \tag{3.36}$$

For example, for the overdamped double-well oscillator with multiplicative and additive noise

$$\frac{dx}{dt} = ax - bx^3 + \xi_1\left(x\right)x + \xi_2\left(x\right) \tag{3.37}$$

one gets [33]

$$\eta = 1 - a\tau_2 + \left(2\tau_1 + 3\tau_2\right)bx^2.$$

Note, that equation (3.35) becomes exact both at $\tau \to 0$ and $\tau \to \infty$, and hence is expected to be a good approximation for intermediate τ as well. In such a way equation (3.29) with colored noise is reduced to equation (3.35) with white noise.

3.3 Brownian motion and anomalous diffusion

As we have seen in the previous section, a Brownian particle subjected to white noise is characterized by the normal diffusion where for long times $\left\langle\left(\Delta x\right)^2\right\rangle$ increases linearly with time. However, it turns out that in many fields of natural science (dozens of examples can be found in the comprehensive reviews [34], [35]) diffusion is anomalous, namely for large t

$$\left\langle\left(\Delta x\right)^2\right\rangle \sim t^\alpha \tag{3.38}$$

with $\alpha \neq 1$. The cases with $\alpha < 1$ and $\alpha > 1$ are called subdiffusion and superdiffusion, respectively.

We show now [36] that the anomalous diffusion can be obtained from the standard Langevin equation (3.1) with the non-white correlator $r(z)$ defined in (2.1). On the substitution of (2.1) into (3.4), one gets for the dispersion of the particle position $\langle x^2(t)\rangle$, with $x(t = 0) = 0$,

$$\langle x^2(t)\rangle = \frac{1}{\gamma m}\int_0^t dz F(z)\left\{2\exp\left[-\frac{\gamma}{m}(t-z)\right]\right.$$
$$\left. -\exp\left[-\frac{\gamma}{m}(2t-z)\right] - \exp\left(-\frac{\gamma}{m}z\right)\right\} \qquad (3.39)$$

where

$$F(z) = \int_0^{|z|} du\, r(u)\,(|z|-u). \qquad (3.40)$$

It turns out [36], that

a) For $0 < F(\infty) < \infty$, (which requires $\lim_{v\to\infty}\int_0^v r(u)du \sim 1/v$) $\langle x^2(t)\rangle$ tends to a finite limit as $t\to\infty$, i.e. the diffusion is confined (stochastic localization).

b) For $F(\infty) = \infty$, the type of diffusion depends on the noise strength $R = \int_0^\infty r(z)\,dz$, namely, $0 < R < \infty$ corresponds to normal diffusion, $R = \infty$ results in superdiffusion, and $R = 0$ (contribution from regions of positive and negative correlations balance) gives subdiffusion.

One can obtain the anomalous diffusion even in the case of white noise in equation (3.1), if the viscosity coefficient η is time-dependent, $\eta = \eta(t)$, say having a form

$$\eta = \frac{a}{t^\beta}. \qquad (3.41)$$

It turns out [37], that for $0 < \beta < 1$ one gets the subdiffusion for $a > 0$ and the stretched exponent $\langle x^2\rangle \approx \exp\left(-\frac{2|a|t^{1-\beta}}{1-\beta}\right)$ for $a < 0$, $\langle x^2\rangle \approx t\ln t$ dependence for $\beta = 1$ and $a < 1/2$, and localization for $\beta < 0$ and $a > 0$.

3.4 Brownian motion near the critical point

The foregoing theory of Brownian motion was based on the small parameter $l/L < 1$, where L is the characteristic size of the Brownian particle, and l is the size of the surrounding molecules. One can consider the opposite limit case, $l/L > 1$, where l is the characteristic size of an inhomogeneity in the medium, such as the size of clusters near the liquid-gas critical point or smallest scale of turbulence in a turbulent flow. The cause of stochasticity is completely different in these two cases. While in the usual Brownian motion the random force is associated with random collisions of molecules with the Brownian particle, in the latter case the stochasticity of the motion of the particle, which we still call the "Brownian particle", is provided by the finite lifetime of clusters and randomness of their decay. The Brownian particle is captured by a cluster, moves for some time inside the given cluster, and after the decay of this cluster the particle is captured by another cluster and so on. For simplicity, we assume [38] that immediately after release by a cluster the particle is captured by another cluster, so in the frame moving with the mean velocity (if such velocity exists) the equation of motion of the particle inside a cluster has the following form

$$\frac{dv}{dt} = -\gamma_0 (v - w) \tag{3.42}$$

where $\gamma_0 = \gamma/m$, and v and w are the velocities of a particle and a cluster, respectively. It is assumed in (3.42) that the drag force acting upon the particle is linear in its velocity relative to that of the cluster at the same point and time. Equation (3.42) has the familiar form (3.1) describing Brownian motion with a random force $\xi(t) = \gamma_0 w$. However, if this noise is not white one has to rewrite equation (3.42) in the form

$$\frac{dv}{dt} - \int_0^t \gamma(t - z) v(z) \, dz = \xi(t) \tag{3.43}$$

and according to the so-called second fluctuation-dissipation theorem [39],

$$\langle \xi(t) \xi(0) \rangle = \langle v^2 \rangle \gamma(t). \tag{3.44}$$

In the case of white noise $\gamma(t) = \gamma_0 \delta(t)$, and (3.43) reverts to (3.42). We would like to compare the diffusion coefficients D of a Brownian particle and ϵ of a medium. These coefficients can be expressed as integrals of the velocity correlation functions [39],

$$D = \int_0^\infty \langle v(0) v(z) \rangle \, dz,$$

$$\epsilon = \int_0^\infty \langle w(0) w(z) \rangle \, dz = \frac{1}{\gamma_0^2} \int_0^\infty \langle \xi(0) \xi(z) \rangle \, dz.$$

(3.45)

On multiplying Eq.(3.43) by $v(0)$, averaging the resulting equation, and taking into account the statistical independence of the velocities of particle and medium, $\langle v(0) w(t) \rangle = 0$, one obtains a solution for the correlation function of a Brownian particle $y(t) \equiv \langle v(t) v(0) \rangle$,

$$y(t) = \langle v^2 \rangle - \int_0^t d\tau \int_0^\tau \gamma(t - z) y(z) \, dz$$

$$= \langle v^2 \rangle - \int_0^t y(s) \, ds \int_s^t \gamma(\tau - s) \, d\tau.$$

(3.46)

The order of integration was changed on passing to the last equality in (3.46). On substituting Eq. (3.44) into (3.46), taking $t \to \infty$, with $y(\infty) = 0$, one derives from (3.46) that

$$\int_0^\infty y(s) \, ds \int_0^\infty \langle w(0) w(z) \rangle \, dz = \langle v^2 \rangle^2.$$

(3.47)

Finally, by using (3.45), we obtain from (3.47) that

$$D = \frac{\langle v^2 \rangle^2}{\gamma_0^2 \epsilon}.$$

(3.48)

One can rewrite Eq.(3.48) using the fluctuation-dissipation theorem (3.44) with $\xi = \gamma_0 w$. Then, the denominator in Eq.(3.48) is reduced to $\langle v^2 \rangle \int_0^\infty \gamma(t) \, dt = \langle v^2 \rangle \gamma(0)$, where $\gamma(0)$ is the Fourier -component of the function $\gamma(t)$ with zero frequency. Finally, equation (3.48) reduces to

$$D = \frac{\langle v^2 \rangle}{\gamma(0)} \tag{3.49}$$

which is formally similar to the Einstein formula for the diffusion coefficient $D = \kappa T/\gamma = \langle v^2 \rangle /\gamma$.

All our previous analysis was based on the fluctuation-dissipation theorem which presumed the interaction and energy exchange between a Brownian particle and surrounding medium leading to the equilibrium. A similar situation takes place in particle diffusion in plasma turbulence, where this interaction is governed by electromagnetic interaction [40]. However, the situation is different in the case of diffusion of a small particle in a turbulent flow, since the fluctuation-dissipation theorem does not apply in this case [41]. As a result [41], in this case $D = \epsilon$.

The problem considered in this section is a special case of the general theory of the two-state random walk in which a particle can be in one of two states for a random period of time, with each of the states having different dynamics and a different switch density for a jump to the second state [42].

Chapter 4

Overdamped harmonic oscillator with additive noise

If one neglects the inertia term in the dynamic equation of a nonlinear oscillator with additive random force $\xi(t)$, one gets the stochastic differential equation of the form

$$\frac{dx}{dt} + f(x) = \xi(t) \tag{4.1}$$

or the linearized equation

$$\frac{dx}{dt} + ax = \xi(t). \tag{4.2}$$

In the case of white noise, one can replace the nonlinear Langevin equation (4.1) by the appropriate Fokker-Planck equation as it was described in section 3.2, while for the colored noise one has to use some approximations such as UCNA considered in section 3.2.3. On the other hand, the linear equation (4.2) can be solved for an arbitrary form of noise. In the absence of random forces, the solution of equation (4.2) has a form $x(t) = x(0)\exp(-at)$, i.e., it vanishes at $t \to \infty$. On the other hand, in the absence of viscosity, $a = 0$, the second moment $\langle x^2 \rangle$ diverges in time. The solution in this case is [36],

$$\langle x^2(t) \rangle = 2F(t) \tag{4.3}$$

where $F(u)$ is defined in (3.40). If $F(\infty) < \infty$, $\langle x^2(t) \rangle$ tends to a finite value at $t \to \infty$. For $F(\infty) = \infty$, $\langle x^2(t) \rangle$ diverges at $t \to \infty$ (stochastic acceleration of the particle by random driving force). For white noise $\xi(t)$, $F(u) = Du$ (Fermi acceleration [43]).

Coming now back to equation (4.2), one can easily obtain the solution of this equation, with zero initial condition $x(0) = 0$,

$$\langle x^2(t)\rangle = \int_0^t dt_1 \int_0^t dt_2 \exp\left[-a\left(2t - t_1 - t_2\right)\right]\langle \xi(t_1)\xi(t_2)\rangle \quad (4.4)$$

which after substitution of (2.1) and simple transformations takes the form (3.39)

$$\langle x^2(t)\rangle = a\int_0^t du F(u)\left[\exp\left(-au\right) - \exp[-a\left(2t - u\right)]\right]$$
$$+ 2F(t)\exp\left(-at\right). \quad (4.5)$$

In general, the correlator between $\xi(t)$ and $\xi(t_1)$ decreases as $|t - t_1|$ increases. Then, the asymptotic behavior of (4.5) is defined by the Laplace transform of $F(u)$,

$$\langle x^2(t)\rangle_{t\to\infty} = a\int_0^\infty du F(u)\exp\left(-au\right). \quad (4.6)$$

4.1   Additive white noise

For additive white noise $\xi(t)$, the Fokker-Planck equation (3.8) associated with the Langevin equation (4.2) has the following form

$$\frac{\partial P}{\partial t} = \frac{\partial}{\partial x}\left(axP\right) + D\frac{\partial^2 P}{\partial x^2}. \quad (4.7)$$

This equation was solved earlier in (3.8)-(3.9) leading to

$$\langle x\rangle = x(0)\exp\left(-at\right),$$
$$\langle x^2\rangle = x(0)^2 \exp\left(-2at\right) + \frac{D}{a}\left[1 - \exp\left(-2at\right)\right], \quad (4.8)$$
$$\sigma^2 = \frac{D}{a}\left[1 - \exp\left(-2at\right)\right].$$

4.2 Additive noise and periodic force

A solution of equation

$$\frac{dx}{dt} + ax = A\cos(\Omega t) + \eta(t) \tag{4.9}$$

can be obtained analogously to (4.7)-(4.8) by replacing ax in (4.7) by $ax - A\cos(\Omega t)$. Then the final results will contain the oscillating function which performs a periodic motion around $x(0)$ in the presence of the periodic force,

$$\overline{\langle x \rangle} = x(0)\exp(-at) + \frac{A}{a^2 + \Omega^2}\left\{a\left[(\cos(\Omega t) - 1)\right] + \Omega\sin(\Omega t)\right\}. \tag{4.10}$$

Chapter 5

Overdamped harmonic oscillator with multiplicative noise

5.1 Multiplicative noise (shift of stable points)

Let us start with the simplest form of multiplicative noise

$$\frac{dx}{dt} = -ax + \xi(t)x. \tag{5.1}$$

As we have already seen, in the absence of noise, the solution of this equation vanishes at $t \to \infty$. The solution of (5.1) is

$$x(t) = \exp\left[-at + \int_0^t \xi(z)\,dz\right] \tag{5.2}$$

and the first moment will have the following forms depending on the type of noise [44]: for white noise,

$$\langle x(t) \rangle = x(0)\exp\left[(-a+D)t\right], \tag{5.3}$$

and for dichotomous noise,

$$\langle x(t) \rangle = x(0)\exp\left(-at\right)\frac{b_1\exp(b_2t) - b_2\exp(b_1t)}{b_1 - b_2}, \tag{5.4}$$

$$b_{1,2} = -\lambda \pm \sqrt{\lambda^2 + \sigma^2}.$$

One sees from (5.3)-(5.4), that, by virtue of multiplicative noise, the first moment is not only not zero, but diverges at $t \to \infty$ for $D > a$, and $\sigma^2 > a^2 + 2a\lambda$, respectively.

From (5.2) one gets for the first moment

$$\langle x(t) \rangle = x(0) \exp\left(-at + \frac{1}{t} \int_0^t \int_0^t \langle \xi(t_1) \xi(t_2) \rangle \, dt_1 dt_2 \right). \quad (5.5)$$

In the limit $t \to \infty$ this equation takes the form

$$\langle x(t) \rangle = x(0) \exp\left[-at + \pi \overline{S}(0)\right] \quad (5.6)$$

where

$$\overline{S}(0) = \lim_{t \to \infty} \frac{1}{t} \int_0^t S(z, 0) \, dz \quad (5.7)$$

is the zero component of the Fourier transform of the correlator (power spectrum),

$$S(t, \omega) = \frac{1}{2\pi} \int_{-\infty}^{\infty} \langle \xi(t+z) \xi(t) \rangle \exp(i\omega z) \, dz. \quad (5.8)$$

These formulas show that the effect of multiplicative (parametric) noise at large times is due to the low-frequency (close to zero) noise power spectrum. This is exactly the opposite of that in the case of the parametric periodic force [45]. Indeed, the stabilization of the upper vertical state of the pendulum is achieved by the high-frequency harmonic vibrations of its suspension [46].

5.2   Multiplicative and additive noises

The simplest generalization of (5.1) involves, along with a multiplicative noise, an additional additive noise,

$$\frac{dx}{dt} = -ax + \sqrt{D}\xi(t)x + \sqrt{\alpha}\eta(t). \quad (5.9)$$

Both types of noise have been intensively studied in different problems. To mention just a few: a quantum dimer [47], a coupled neutron network [48], synchronization of chaotic oscillators [49], detecting the gravitational background [50], and medical imaging [51].

5.2.1 *Two white noises*

Under the simplest assumption that two sources of noise in (5.9) are uncorrelated having the equal strength,

$$\langle \xi(t_1)\xi(t_2)\rangle = \langle \eta(t_1)\eta(t_2)\rangle = 2\delta(t_1 - t_2), \tag{5.10}$$

one can get the exact solution [52] even for the case when the coefficient a is the arbitrary function of time, $a = a(t)$.

The solution of equation (5.9) is given by

$$x(t) = K(t)L(t)\left[x(t=0) + \sqrt{\alpha}\int_0^t \frac{\eta(t_1)\,dt_1}{K(t_1)L(t_1)}\right] \tag{5.11}$$

where

$$K(t) = \exp\left[-\int_0^t a(t_1)\,dt_1\right]; \quad L(t) = \exp\left[\sqrt{D}\int_0^t \xi(t_1)\,dt_1\right]. \tag{5.12}$$

For $a = const$, the first two average moments are

$$\langle x(t)\rangle = x(t=0)\exp\left[-(a-D)t\right],$$

$$\langle x^2(t)\rangle = x^2(t=0)\exp\left[-2(a-2D)t\right] \tag{5.13}$$

$$+ \frac{\alpha}{a-2D}\left[1 - \exp\left[-2(a-2D)t\right]\right].$$

Two limit cases can be obtained from (5.13). For $\alpha = 0$ and $a = const$, one gets the moments of a system without additive noise, and for $D = 0$, those without multiplicative noise.

5.2.2 *Two correlated white noises*

Let us consider equation (5.9) with $a = const$ and two sources of white noise (5.10) which are correlated,

$$\langle \xi (t_1) \eta (t_2) \rangle = 2\kappa \sqrt{D\alpha} \delta (t_1 - t_2). \tag{5.14}$$

Correlations between different noise sources may occur when they both have the same origin, as in laser dynamics [53], or when strong noise leads to an appreciated change in the internal structure of a system and hence in its internal noise. There are few simple rules for obtaining the solution of a problem with correlated multiplicative and additive noise in terms of the results with non-correlated noise [54].

It turns out that for delta-correlated noise (5.14),

$$\langle x (t) \rangle = x (t{=}0) \exp \left[- (a{-}D) t\right] + \frac{\kappa \sqrt{D\alpha}}{a{-}D} \{1{-}\exp \left[- (a{-}D) t\right]\},$$

$$\langle x^2 (t) \rangle = \frac{\alpha}{a{-}2D} + \left[x^2 (t{=}0) - \frac{\alpha}{a{-}2D} \right] \exp \left[-2 (a{-}2D) t\right] \tag{5.15}$$

$$+ 6\kappa \sqrt{D\alpha} \int_0^t \langle x (u) \rangle \exp \left[-2 (a{-}2D) u\right] du$$

which for $\kappa = 0$ reduces to (5.13). The stationary $(t \to \infty)$ correlation function is given by

$$\langle x (t) x (t + \tau) \rangle_{t \to \infty}$$

$$= \frac{2D}{a - 4\alpha} + \frac{4\kappa^2 D\alpha}{(a - 2\alpha)^2} + \frac{8\kappa^2 \alpha D (a - \alpha)}{(a - 2\alpha)^2 (a - 4\alpha)} \exp \left[- (a - 2\alpha) |\tau|\right]. \tag{5.16}$$

5.2.3   *Two color dichotomous correlated noises*

We do not bring here the appropriate cumbersome formulae which can be found in [55].

5.3 Multiplicative color noise and periodic signal (stochastic resonance (SR))

The interesting phenomenon of stochastic resonance may appear in a system subject to both random and periodic force. Manifestation of SR is typically related to $x(t)$ at the characteristic frequency and can be expressed in terms of different functions that are calculated in terms of $x(t)$ such as the autocorrelation function, the power spectrum, or a signal-to-noise ratio which behave non-monotonically as a function of the noise amplitude, or the correlation rate.

It first seemed [56] that all three ingredients - nonlinearity, periodic and random forces - are necessary for the onset of SR. However, it later became clear that SR may appear without a random force (replaced by a chaotic signal [57]), without a periodic force (autonomous SR [58], aperiodic SR [59]) or by replacing the characteristic frequency by some fluctuation rate [60]), and in linear systems (with multiplicative noise [61], [62]). The latter has been shown for the first-order differential equation describing an overdamped oscillator with multiplicative noise (Ornstein-Uhlenbeck [62], Gaussian [63], Poisson [64], or composite [65] noise).

Since the nonlinearity presents difficulties for the theoretical analysis, the linear models with multiplicative noise are of special interest. These models on the one hand show quasi-nonlinear behavior including stochastic resonance [61], [62], and on the other hand, allow an exact analytical treatment.

On addition of a periodic force to equation (5.1) one gets

$$\frac{dx}{dt} = -ax + \xi(t)x + A\sin(\Omega t). \tag{5.17}$$

The averaged solution of (5.17) has the following form

$$\langle x(t)\rangle = \exp(-at)\left\{x(0)\left\langle\exp\left(\int_0^t \xi(z)\,dz\right)\right\rangle\right.$$

$$\left. + A\int_0^t \sin(\Omega u)\exp(au)\left\langle\exp\left[\int_0^{t-u}\xi(z)\,dz\right]\right\rangle\right\}. \tag{5.18}$$

Differentiating $\left\langle \exp\left(\int_0^t \xi\left(z\right) dz\right)\right\rangle$ one gets

$$\frac{d}{dt}\left\langle \exp\left(\int_0^t \xi\left(z\right) dz\right)\right\rangle$$

$$= \left\langle \xi\left(t\right) \exp\left[\int_0^t \xi\left(z\right) dz\right]\right\rangle$$

$$= \left\langle \xi\left(t\right) \left[1 + \int_0^t \xi\left(z\right) dz + \ldots\right]\right\rangle$$

$$= \int_0^t \left\langle \xi\left(t\right) \xi\left(u\right) \exp\left(-\int_0^u \xi\left(z\right) dz\right)\right\rangle du$$

$$\simeq \int_0^t \left\langle \xi\left(t\right) \xi\left(u\right)\right\rangle \left\langle \exp\left(\int_0^u \xi\left(z\right) dz\right)\right\rangle du. \qquad (5.19)$$

The splitting of the average of product of three random forces in the last line in (5.19) is exact for white and dichotomous noises, and it is a quite good approximation for other types of colored noise.

Using the Laplace transform to solve equation (5.19), and substituting this solution into (5.18), one gets

$$\langle x\left(t\right)\rangle = \left[x\left(0\right) + \frac{A\Omega}{\left(a - b_1\right)^2 + \Omega^2}\right]\left(\frac{b_1}{b_1 - b_2}\right)\exp\left[\left(b_2 - a\right) t\right]$$

$$+ \left[x\left(0\right) + \frac{A\Omega}{\left(a - b_2\right)^2 + \Omega^2}\right]\left(\frac{b_2}{b_2 - b_1}\right)\exp\left[\left(b_1 - a\right) t\right]$$

$$+ A\left[\frac{\Omega^2 + \left(a + \lambda\right)^2}{\left[\left(a - b_1\right)^2 + \Omega^2\right]\left[\left(a - b_2\right)^2 + \Omega^2\right]}\right]^{1/2}\cos\left(\Omega t + \phi\right)$$

$$(5.20)$$

where $b_{1,2}$ were defined in (5.4).

For white noise equation (5.20) reduces to

$$\langle x(t) \rangle = \left[x(0) + \frac{A\Omega}{(a-D)^2 + \Omega^2} \right]_1 \exp\left[-(a-D)t \right]$$

$$+ \frac{A}{\left[(a-D)^2 + \Omega^2 \right]^{1/2}} \cos(\Omega t + \phi). \qquad (5.21)$$

For all $a > D$, where $\langle x(t) \rangle$ remains bounded, the amplitude of the stationary solution in equation (5.21) is a monotonic function of D reaching maximum at $D = a$ showing some SR-like behavior [66].

The average stationary solution of equation (5.17) can be obtained from the limit of (5.20) at $t \to \infty$ which gives

$$\langle x(t) \rangle_{st} = \sqrt{ \frac{\Omega^2 + (a+\lambda)^2}{\Omega^4 + 2(a^2 + 2a\lambda + 2\lambda^2 + \sigma^2)\Omega^2 + (a^2 + 2a\lambda - \sigma^2)^2} }$$

$$\times A\cos(\Omega t + \phi). \qquad (5.22)$$

The amplitude of the stationary solution depends on the dynamic parameter a, the amplitude A and the frequency Ω of the periodic force and the strength σ^2 and the correlation rate λ of the noise. Some typical graphs displayed in Figs. 5.1 and 5.2 show the non-monotonic dependence of $\langle x(t) \rangle_{st}$ as a function of σ^2 and λ (stochastic resonance). Analysis of (5.22) shows that the maxima occurs at

$$\sigma_{max}^2 = a^2 + 2a\lambda - \Omega^2. \qquad (5.23)$$

The appearance of SR for a process described by equation (5.17) with dichotomous noise $\xi(t)$ can be understood in the following way. For the limiting case $a = \xi = 0$, a particle executes a periodic motion with an amplitude A/Ω. If there is no random force, $\xi = 0$, but $a \neq 0$, a particle will move along the parabola $U = ax^2/2$. For dichotomous noise $\pm\sigma$, a particle moves along the parabola $U = (a+\sigma)x^2/2$, then jumps at rate τ to the parabola $U = (a-\sigma)x^2/2$, etc. For $\sigma > a$, but $\sigma < (a^2 + 2a\lambda)^{1/2}$, these two parabolas have curvatures of opposite signs, and they act in opposite directions tending to increase (decrease) the displacement x of a particle. Their mutual influence is defined by noise which causes

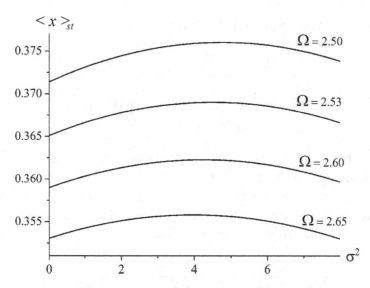

Fig. 5.1   The amplitude of a stationary signal as a function of the noise strength for $a = A = 1$, $\lambda = 10$ and different frequencies of the periodic field. [Reprinted from Ref. 62 with permission from EDP Sciences.]

jumps between the parabolas and by a periodic force which determines the amplitude of oscillations along the parabolas. Accordingly, the amplitude of the stationary signal, Eq. (5.22), has a maximum as a function of the noise strength.

Contrary to the dichotomous noise, the white noise has no characteristic frequency so there is no "adjustment" for the frequency Ω of an external field, and there is no stochastic resonance for white noise.

The more general case of additive and multiplicative colored noise as well as a periodic signal has been considered in [55].

A useful generalization of equation (5.9) has been performed in [67] by adding the periodicaly-modulated noise term of the form $A\cos(\Omega t)\eta(t)$. Both $\xi(t)$ and $\eta(t)$ were taken as asymmetric dichotomous noises. The signal-to-noise ratio turns out to be a non-monotonic function of each one of the following quantities: strength of noise, correlation time of noise and the frequency of the periodic field.

Fig. 5.2 The amplitude of a stationary signal as a function of the correlation time for $a = A = \sigma^2 = 1$ and different frequencies of the periodic field. [Reprinted from Ref. 62 with permission from EDP Sciences.]

5.4 Stochastic resonance in a overdamped system with signal-modulated noise

In the previous section we considered an overdamped oscillator subject to multiplicative noise and an additive periodic signal. However, one can also consider the case when the signal is multiplied by noise. Such a problem has been considered in [68] for bistable potentials in connection with the periodically modulated noise arising in some optical and astronomical devices. Recently, a similar problem has been considered for linear systems [69]. We bring here the results of the latter article.

The overdamped oscillator is subject to signal-modulated noise $\xi(t)$ and additive noise $\eta(t)$,

$$\frac{dx}{dt} + ax = \xi(t)\left[1 + A\cos(\Omega t)\right] + \eta(t) \qquad (5.24)$$

where $\xi(t)$ and $\eta(t)$ are exponentially correlated color noises

$$\langle \xi(t) \rangle = \langle \eta(t) \rangle = 0;$$

$$\langle \xi(t_1) \xi(t_2) \rangle = \frac{Q}{\tau} \exp\left[-\frac{|t_1 - t_2|}{\tau}\right],$$

$$\langle \eta(t_1) \eta(t_2) \rangle = \frac{D}{\tau} \exp\left[-\frac{|t_1 - t_2|}{\tau}\right], \qquad (5.25)$$

$$\langle \xi(t_1) \eta(t_2) \rangle = 2\kappa \frac{\sqrt{DQ}}{\tau} \exp\left[-\frac{|t_1 - t_2|}{\tau}\right].$$

The solution of equation (5.24) has the following form

$$x(t) = \exp(-at)\left\{ x(t=0) + \int_0^t du \exp(au) \xi(u) \left[1 + A\cos(\Omega u)\right] \right.$$

$$\left. + \int_0^t du \exp(au) \eta(u) \right\}. \qquad (5.26)$$

From equation (5.26) one gets for the stationary value of the power spectrum of the correlation function,

$$S(\omega) = \int_{-\infty}^{\infty} du \exp(i\omega u) \langle x(t+u) x(t) \rangle_{t\to\infty} = S_{\text{signal}} + S_{\text{noise}} \qquad (5.27)$$

with the signal-to-noise ratio

$$R = \frac{\displaystyle\int_{-\infty}^{\infty} S_{\text{signal}}}{\displaystyle\int_{-\infty}^{\infty} S_{\text{noise}}}. \qquad (5.28)$$

The slightly cumbersome result of this exact calculation is [69]

$$R = \frac{aQA^2 a_+}{Q + D + 2\kappa\sqrt{QD}}$$

$$\times \left[\frac{a_+ a_- + \Omega a_+}{2\left(a_-^2 + \Omega^2\right)\left(a_+^2 + \Omega^2\right)} + \frac{a_+}{4a\left(a_+^2 + \Omega^2\right)} - \frac{a_-}{4a\left(a_-^2 + \Omega^2\right)} \right] \qquad (5.29)$$

with $a_\pm = a \pm \tau^{-1}$.

Analysis of equation (5.29) shows the existence of stochastic resonance: for the negative correlation region $-1 \leq \kappa < 0$, one finds a non-monotonic dependence of the signal-to-noise ratio R on both modulated noise strength Q and non-modulated noise strength D. It is remarkable that both effects remain even in the white noise limit, $\tau \to 0$. In addition, one gets a non-monotonic dependence of R on both the correlation rate τ^{-1} and the frequency Ω of the modulated field.

One additional comment is necessary [70]. The stochastic resonance phenomenon (5.29) for negative correlation which occurs for an arbitrary form of correlation functions in (5.25), exists already in the input signal (5.24) for the chosen forms of noise.

Chapter 6

Overdamped single-well oscillator

For the sake of simplicity, we start from the simplest nonlinear oscillator model described by the following equation,

$$\frac{dx}{dt} = ax - bx^2 \tag{6.1}$$

with growth rate a and decay rate b.

The solution of equation (6.1) is

$$x(t) = \frac{ax(t=0)}{bx(t=0) - [bx(t=0) - a]\exp(-at)} \tag{6.2}$$

which shows that the stability is defined by the initial position $x(t=0)$: the motion is stable for $x(t=0) > 0$, and it is unstable $(x(t) \to \infty)$ for $x(t=0) < 0$.

For obvious reasons Eq. (6.1) is called the birth-death equation for the growth of population, cells, etc., where x is assumed to be non-negative, $x \geq 0$. This equation has two stable points, $x_1 = 0$ and $x_2 = a/b$. The stability of these solutions is defined by linearizing equation (6.1) around the stable points. This simple analysis shows that $x_1 = 0$ is stable for $a < 0$, whereas $x_2 = a/b$ is stable for $a > 0$. Let us now include additive noise into equation (6.1), and the fluctuations of parameters a and b leading to multiplicative noise, starting with the fluctuations of the parameter a.

41

6.1 Steady state

6.1.1 *White noises*

Introducing correlated white noise in (6.1), yields [71]

$$\frac{dx}{dt} = ax + x\xi(t) - bx^2 + \eta(t) \tag{6.3}$$

with

$$\langle \xi(t_1)\xi(t_2) \rangle = 2D\delta(t_1 - t_2); \quad \langle \eta(t_1)\eta(t_2) \rangle = 2\alpha\delta(t_1 - t_2),$$
$$\langle \xi(t_1)\eta(t_2) \rangle = 2\kappa\sqrt{D\alpha}\delta(t_1 - t_2). \tag{6.4}$$

For a system subject only to multiplicative noise ($\eta = 0$ in (6.4)), there are [72] two critical values of the control parameter: $a = 0$ and $a = D$. For $a < 0$, the fixed point $x = 0$ is stable, and the stationary distribution function $P_{st}(x)$ is concentrated at zero as a delta-function, $P_{st}(x) \sim \delta(x)$. For $a > 0$, the fixed point $x = 0$ becomes unstable, but remains the most probable for $0 < a < D$. Finally, for $a > D$, the divergence at zero disappears and the point $x = a/b$ becomes stable.

The Fokker-Planck equation, corresponding to the Langevin equation (6.3) with noise (6.4) has the form [71]

$$\frac{\partial P}{\partial t} = -\frac{\partial}{\partial x}[A(x)P] + \frac{\partial^2}{\partial x^2}[B(x)P] \tag{6.5}$$

where the drift term $A(x)$ and the diffusive term $B(x)$ are

$$A(x) = ax - bx^2 + Dx - \kappa\sqrt{D\alpha},$$
$$B(x) = Dx^2 - 2\kappa\sqrt{D\alpha}x + \alpha. \tag{6.6}$$

The time-dependent equation (6.5) is quite complicated, but its stationary ($t \to \infty$) solution can be easily found. For $0 \leq \kappa < 1$, one gets

$$P_{st}(x) = NB(x)^{\frac{a}{2D} - \frac{1}{2} - \sqrt{\frac{ab^2\kappa^2}{D^3}}}$$

$$\times \exp\left[-\frac{bx}{D} + \frac{E}{\sqrt{D\alpha(1-\kappa^2)}} \arctan\left(\frac{Dx - \kappa\sqrt{D\alpha}}{\sqrt{D\alpha(1-\kappa^2)}} \right) \right]$$

(6.7)

where N is the normalization factor, and

$$E = \frac{b\alpha}{D} - \left[\kappa\sqrt{\frac{\alpha}{D}}\left(a + 2\kappa\sqrt{\frac{\alpha b^2}{D}} \right) \right].$$

(6.8)

The question arises (as in section 5.1) of the shift of the deterministic fixed points $x_1 = 0$, and $x_2 = a/b$ due to the presence of noise. As follows from (6.7), the extrema of $P_{st}(x)$ obey the following equation

$$bx^2 + (D - a)x - \kappa\sqrt{D\alpha} = 0.$$

(6.9)

The solutions of this equation for small correlation between noise κ are

$$x_1 \approx \frac{\kappa\sqrt{D\alpha}}{D-a}, \quad x_2 \approx \frac{a}{b} - \frac{D}{b} - \frac{\kappa\sqrt{D\alpha}}{D-a}.$$

(6.10)

Analysis of the influence of noise on $P_{st}(x)$ yields the following results [71]:

1. As the value of correlation parameter κ increases, $P_{st}(x)$ increases at small x and decreases at large x.

2. As the strength of the additive noise α increases, the maximum value of a peak increases at small values of x and decreases at large values of x. The peak is flattened and almost disappears at large values of α (effect of diffusion).

3. As the strength of the multiplicative noise D increases, the maximum of $P_{st}(x)$ moves to smaller values of x (effect of the drift term).

6.1.2   *Multiplicative noise (gene selection)*

A slightly different dimensionless form of equation (6.3) has been used [73] to describe gene selection, namely,

$$\frac{dx}{dt} = a_1 - a_2 x + \lambda x (1 - x) + \xi(t) x (1 - x). \qquad (6.11)$$

Special values of the parameters a_1 and a_2 have been chosen in the standard gene model, $a_1 = 1/2$, $a_2 = 1$. For white noise $\xi(t)$, the Fokker-Planck equation corresponding to (6.11) has the following form

$$\frac{\partial P(x,t)}{\partial t} = \left\{ -\frac{\partial}{\partial x} \left[\frac{1}{2} - x + \lambda x (1 - x) \right] \right.$$

$$\left. + D \frac{\partial}{\partial x} x (1 - x) \frac{\partial}{\partial x} x (1 - x) \right\} P(x,t) \qquad (6.12)$$

with the stationary solution

$$P_{st}(x) = N (1 - x)^{-1 - \frac{\lambda}{D}} x^{-1 + \frac{\lambda}{D}} \exp \left[-\frac{1}{2Dx(1 - x)} \right]. \qquad (6.13)$$

For the symmetric case, $\lambda = 0$, $P_{st}(x)$ has only one maximum at $x = 1/2$ for $0 < x < 1$ if $D < 2$. But if $D > 2$, this point becomes a minimum and two maxima occur at

$$x = \frac{1}{2} \left(1 \pm \sqrt{1 - \frac{2}{D}} \right). \qquad (6.14)$$

The value $D = 2$ defines the noise-induced phase transition analogous to equilibrium phase transitions [74]. For $\lambda \neq 0$, the qualitative picture is the same, but the transition occurs at $D = D(\lambda)$.

Finally, for Ornstein-Uhlenbeck noise $\xi(t)$, quite complicate approximate calculations show [75] that the transition described above occurs at $D = 2(1 - \tau)$. However, the increase of the dimensionless correlation rate τ may lead to a unimodal-bimodal transition even for $D < 2$.

6.1.3 *Dichotomous noise*

For dichotomous multiplicative noise where $\xi(t)$ takes two values $\pm\sigma$ with a rate of jumps $\lambda/2$ between these two states, one can solve equation (6.3) for general forms of the deterministic part. For simplicity, consider the case of only multiplicative noise,

$$\frac{dx}{dt} = f(x) + g(x)\xi(t) \tag{6.15}$$

with $f(x)$ and $g(x)$ being arbitrary nonlinear functions of x, and $\xi(t)$ is dichotomous noise. The Fokker-Planck equations for the distribution functions $P(x,t,\sigma)$ and $P(x,t,\sigma)$ equivalent to the Langevin equation (6.15) have the following form [76]

$$
\begin{aligned}
\frac{\partial P(x,t,\sigma)}{\partial t} &= -\frac{\partial}{\partial x}\{[f(x) + g(x)\sigma]P(x,t,\sigma)\} \\
&\quad - \frac{\lambda}{2}[P(x,t,\sigma) - P(x,t,-\sigma)], \\
\frac{\partial P(x,t,-\sigma)}{\partial t} &= -\frac{\partial}{\partial x}\{[f(x) - g(x)\sigma]P(x,t,\sigma)\} \\
&\quad + \frac{\lambda}{2}[P(x,t,\sigma) - P(x,t,-\sigma)].
\end{aligned}
\tag{6.16}
$$

The stationary ($t \to \infty$) solution for the probability density $P_{st}(x) = P(x,t,\sigma) + P(x,t,-\sigma)$ is easily obtained from (6.16)

$$P_{st}(x) = N\frac{|g(x)|}{D_{eff}(x)}\exp\left[\int^x \frac{f(y)}{D_{eff}(y)}dy\right] \tag{6.17}$$

where $D_{eff}(x)$ is an "effective diffusion coefficient" given by

$$D_{eff}(x) = \frac{\sigma^2}{2\lambda}\left[g(x) + \frac{f(x)}{\sigma}\right]\left[g(x) - \frac{f(x)}{\sigma}\right] \tag{6.18}$$

which has to be positive for the stable solution of equation (6.15).

The extrema x_e of $P_{st}(x)$ are the solutions of the following equation

$$f - \frac{\sigma^2}{2\lambda} g \frac{dg}{dx} + \frac{1}{\lambda}\left(2f\frac{df}{dx} - \frac{f^2}{g}\frac{dg}{dx}\right)_{x=x_e} = 0. \qquad (6.19)$$

For equation (6.3), $f(x) = ax - bx^2$ and $g(x) = x$. Therefore, the general form (6.17) of $P_{st}(x)$ reduces to [77]

$$P_{st}(x) = N \, x^{\frac{\lambda a}{\sigma^2 - a^2} - 1} \, |bx + \sigma - a|^{-\frac{\lambda}{2(\sigma - a)} - 1} \, |bx - \sigma - a|^{-\frac{\lambda}{2(\sigma + a)} - 1}, \qquad (6.20)$$

and the equation for the extrema is

$$bx - a + \frac{\sigma^2}{2\lambda} - \frac{2}{\lambda}(a - bx)(a - 2bx) + \frac{1}{\lambda}\left(ax - bx^2\right)^2 = 0. \qquad (6.21)$$

When passing from dichotomous to white noise ($\sigma^2 \to \infty$, $\lambda \to \infty$, at $\sigma^2/\lambda = 2D$), equation (6.21) reduces to (6.9) with $\kappa = 0$, as it should.

6.1.4 *Poisson white noise*

After dichotomous noise, we consider the more realistic form of colored noise, Poisson white noise, defined in (2.11)-(2.12). Both dichotomous and Poisson noise differ from white noise in that they both may fluctuate remaining restricted. This fact is of special importance for the decay rate b in (6.1), which has to be positive in order to get stable solutions to equation (6.1). Here, we bring the results of the appropriate calculations [78].

Consider first the fluctuations of the growth rate: $a(t) = a_0 + \xi(t)$. For $\lambda w < a_0$, $a(t)$ remains positive and the stationary probability distribution

$$P_{st}(x) = N(x - x_1)^{\frac{a_0}{wbx_1} - \frac{1}{w} - 1} x^{-1 - \frac{a_0}{wbx_1}}, \quad x_1 = \frac{a_0 - \lambda w}{b} \qquad (6.22)$$

is defined in the domain (x_1, ∞).

For $\lambda w > a$, $a(t)$ can take negative values, and the probability density

$$P_{st}(x) = N_1 x^{\frac{a_0}{wbx_1} - 1} (x + x_1)^{-a_0}, \quad x_1 = \frac{\lambda w - a_0}{b} \qquad (6.23)$$

is defined in $(0, \infty)$.

Consider now the fluctuations of the decay rate: $b(t) = b_0 + \xi(t)$. For $b_0 > \lambda w$, $b(t)$ is always positive and the stationary probability distribution is

$$P_{st}(x) = N \exp\left(-\frac{1}{wx}\right) x^{\frac{a - b_0 x_1}{wax_1} - 1} (x_1 - x)^{\frac{b_0 x_1 - a}{wax_1}}, \quad x_1 = \frac{a}{b_0 - \lambda w}. \tag{6.24}$$

For $b_0 < \lambda w$, the situation is more complicated. Although the formal solution exists in the region $(0, \infty)$, it tends to infinity as $1/x^2$ at $x \to 0$, which has no physical meaning. Then, fluctuations in the parameter b with the Stratonovich interpretation of multiplicative noise lead to instability. The Ito interpretation produces a different result [79].

6.2 Response to a periodic force (noise-enhanced stability)

In the previous section, we considered the influence of multiplicative noise on equation (6.1). Here we analyze the influence of a periodic force and the additive noise which in the birth-death model represents a periodical and random influence of an environment [80].

Equation (6.1) with force $F \equiv -\frac{dU}{dx} = ax - bx^2$ describes the (overdamped) motion of a particle in the potential $U = -\frac{ax^2}{2} + \frac{bx^3}{3}$. This potential is characterized by two equilibrium points, one metastable, $x = a/b$, and one unstable, $x = 0$, as illustrated in Fig. 6.1. In the presence of noise, a particle always escapes from the potential well, and we are interested in the effect of noise of strength D on the mean escape time in the presence of an external periodic force. Let us start, however, from the deterministic equation

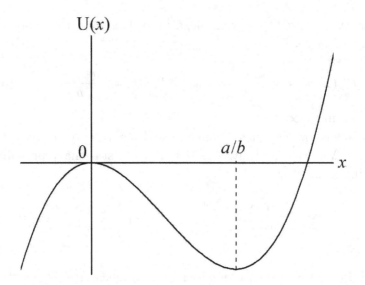

Fig. 6.1   A single-well potential with an unstable point at $x = 0$ and a metastable point at $x = a/b$. [Reprinted figure with permission from Ref. 80. Copyright (1992) by the American Physical Society.]

obtained from (6.1) by adding a periodic force

$$\frac{dx}{dt} = ax - bx^2 + A\cos(\Omega t). \tag{6.25}$$

Although a full solution of this equation can be found in terms of Mathieu functions, we restrict ourselves to a qualitative analysis. An escape from the potential well depends on the initial position of a particle $x(t = 0)$, the amplitude A, and the frequency Ω of an external field. If A is smaller than the barrier height, $A < a^3/6b^2$, the particle will never leave the well provided that the initial position satisfies

$$x(t = 0) \geq \frac{a}{2b}\left(1 - \sqrt{1 + \frac{4Ab}{a^2}}\right). \tag{6.26}$$

This follows from the fact that the upper bound of A ensures that for a fixed t, the right hand side of equation (6.25) has two real roots, which is equivalent to the statement that the particle will

Fig. 6.2 Numerical solution of the dimensionless equation (6.25) for parameters $A = 0.3$, and frequencies a: $\Omega = 0.13196$, b: $\Omega = 0.13197$, c: $\Omega = 0.13198$, and d: $\Omega = 0.13199$. [Reprinted figure with permission from Ref. 80. Copyright (1992) by the American Physical Society.]

remain trapped between the two values of x that correspond to these roots.

It is intuitively clear that the larger the amplitude of the field, the easier it is for the particle to escape from the well, but the extreme sensitivity of the escape to the frequency Ω of the field is much less obvious. Figure 6.2 shows curves $x(t)$ for several values of the frequency. These curves illustrate the fact that escape can occur at the end of any integer number of the cycle time $T = 2\pi/\Omega$ associated with the periodic term in (6.25). The results are extremely sensitive to changes in the frequency. Indeed, a change of only 10^{-6} in the frequency determines whether the particle will escape after the first cycle. In Fig. 6.3, we show the joint action of A and T on the escape from the well. For all values of A and T located above the upper curve, escape occurs even before the end of the first cycle. The uppermost curve corresponds to systems in which the particle escapes at the end of the first cycle. The curve just below corresponds to

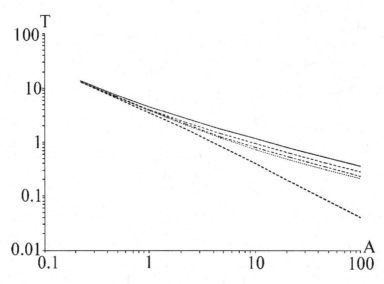

Fig. 6.3   Separatrices in the (A, T) plane. The region above the uppermost curve corresponds to the region in which a particle "escapes" before the end of the first cycle. The region below the lowest curve shows the states in which a particle never escapes, and the intermediate region corresponds to states in which a particle escapes after a finite numbers of cycles. [Reprinted figure with permission from Ref. 80. Copyright (1992) by the American Physical Society.]

systems in which the particle escapes just at the end of the second cycle, etc. The points below the lowest curve in Fig. 6.3 correspond to systems in which escape does not occur.

Let us now consider equation (6.1) supplemented with a periodic force and additive noise

$$\frac{dx}{dt} = ax - bx^2 + A\cos(\Omega t) + \xi(t). \tag{6.27}$$

The result of the numerical solution of equation (6.27) for the escape time $\langle \tau \rangle$ is shown in Fig. 6.4 for the initial condition $x(t = 0) = 1$. These graphs look quite contra-intuitive. Indeed, at not too small A, both periodic force and noise favour escape, and it is believed that these two factors act together. However, as one can see from Fig. 6.4, adding noise to a periodically driven system will increase (!) the escape time for some noise strength. This effect has been named

Fig. 6.4 The average escape time as a function of the logarithm of the noise strength for $A = 0.3$, and the same frequencies as used in Fig. 6.2. [Reprinted figure with permission from Ref. 80. Copyright (1992) by the American Physical Society.]

[81] noise-enhanced stability (NES)[1]. Noise-enhanced stability has now been found experimentally [81].

6.3 Piece-wise model of a metastable state

The effect of noise-enhanced stability (NES) can be explained quantitatively [82] by the example of a piece-wise linear potential $U(x)$ for $x > 0$ (Fig. 6.5)

$$U(x) = \begin{cases} 0, & 0 < x < L \\ E - k(x - L), & L < x < b \end{cases} \qquad (6.28)$$

---

[1] In Hebrew "ness" means "miracle". As is well-known, to live in Palestine and perform miracles is quite dangerous... . Another dangerous prediction is that the addition of a periodic force (wars and the after-war children boom ?) to the birth-death equation, may, at some resonance conditions, increase the time it takes to reach population extinction $x = 0$; i.e., it may be useful for the survival of the species.

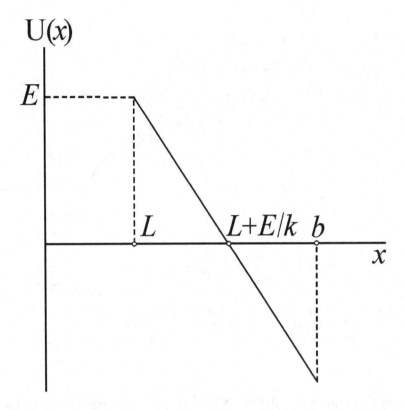

Fig. 6.5   The triangle piece-wise potential described by equation (6.28).

where E is the height of the potential barrier and b is the absorbing boundary. The states at $0 < x < L$ are metastable, and those at $L < x < b$ are unstable. The overdamped periodically driven motion of a particle in the potential (6.28) is described by the following equation

$$\frac{dx}{dt} = -\frac{dU}{dx} + A\sin(\Omega t) + \xi(t) \qquad (6.29)$$

with white noise $\xi(t)$ of strength D.

Consider first the time-independent potential ($A = 0$). If the initial position of a particle $x_0 = x(t = 0)$ is unstable, $L < x_0 < L + E/k$, the average escape time grows in the presence of noise since the particle may jump into the potential well. For very weak noise, the probability of such jumps is very low. Just due to these jumps, the particle will be trapped in the well for a long time (noise enhanced stability).

Let us turn to the case $A \neq 0$. For $x_0 = 0$, the particle at $0 < x < L$ will move according to the equation $x(t) = \frac{A}{\Omega} [1 - \cos(\Omega t)]$. If $2A/\Omega < L$, the particle will always remain inside the region $(0, L)$. However, if $2A/\Omega > L$, the particle surmounts the region $(0, L)$, and its position will change with time as

$$
x(t) = \begin{cases} \dfrac{A}{\Omega} [1 - \cos(\Omega t)], & 0 < t < t_1 \quad 0 < x(t) < L, \\ k(t - t_1) + \dfrac{A}{\Omega} [1 - \cos(\Omega t)], & t > t_1 \quad L < x < b \end{cases}
$$

(6.30)

where t_1 is the time at which the particle crosses the point $x = L$,

$$
t_1 = \frac{1}{\Omega} \arccos\left(1 - \frac{\Omega L}{A}\right).
$$

(6.31)

NES occurs if at time $t_2 = T/2 = \pi/\Omega$, when the periodic force changes its sign, tending to return the particle to the region $(0, L)$, the particle is still located inside the interval $(0, b)$, $x(t_2) < b$. Using (6.31) and the second formula in (6.30), the latter inequality can be rewritten as

$$
\frac{2A}{b} + \frac{k}{b}\left[\pi - \arccos\left(1 - \frac{\Omega L}{A}\right)\right] < \Omega
$$

(6.32)

which, in addition to the previously used inequality

$$
\Omega < \frac{2A}{L},
$$

(6.33)

defines the conditions for the appearance of NES. Just these two inequalities define the regions in the $A - T$ plane explained previously and shown in Fig. 6.3.

The onset of NES has been considered in [82] also in circumstances where a periodic force in (6.28) has been replaced by multiplicative dichotomous noise.

6.4   Rectangular potential barrier (stabilization of metastable state)

In the previous section, we considered the influence of noise on the escape time of a particle in a metastable state, and found that noise may enhance the stability of a particle which otherwise will leave a metastable state under the action of a periodic signal. Here we consider a slightly different situation involving a particle driven by noise from its metastable position, showing that an additional periodic signal is able (like noise in the previous chapter) to increase the stability of a metastable state[2].

As is well known [46], a simple pendulum is stable (metastable) in the vertically downward (upward) position. One can, however, stabilize a metastable position by applying high-frequency parametric oscillations of a pendulum (the "Kapitza pendulum"). As was shown both numerically [83] and analytically [84], the "dynamic stabilization" of a pendulum can be also achieved by adding an additive, and not multiplicative, periodic field. Incidentally, not only does the metastable state becomes stable, but also the stable states become unstable[3].

The question arises how general is this phenomenon, and whether one can stabilize other metastable states as well. In order to perform analytical calculations, we consider in this section the simplest model of a rectangular potential shown in Fig. 6.6. As was shown in the last 10-20 years, many fundamental properties of a particle moving in a nonlinear potential are generic, and not too sensitive to the details of the potential. Therefore, it is worthwhile to consider the simplest potential that allows an analytical solution, in addition to numerical simulations for more realistic potentials. Our model (Fig. 6.6)

---

[2] "Rich become richer" - this is the main slogan of the theory of self-organized criticality in growing networks, very popular nowadays, where a new site prefers to be connected with sites which already have many connections and become thereby even "richer". In this section we are aware of metastable states, which are "poor" - less stable and containing fewer particles than stable states. Our aim is to find what are the possibilities for stabilization of metastable states, how to make them "richer". Our main slogan "poor become richer" is probably less realistic, but certainly more democratic.

[3] The previously declared slogan "poor become richer" is now complemented by the even more revolutionary slogan "rich become poorer".

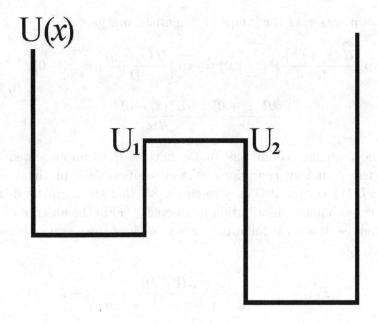

Fig. 6.6   The simplest form of the rectangular piece-wise bistable potential.

involves a particle subject to white noise of strength D moving in a piecewise double-well potential restricted by reflecting walls. The potential barriers U_1 and U_2 are different for the right (stable) and the left (metastable) states.

For the potential $U(x)$ shown in Fig. 6.6, $\frac{dU}{dx} = 0$ everywhere except at the matching points. The Fokker-Planck equation, which can be written in the form

$$\frac{\partial P}{\partial t} = \frac{\partial}{\partial x}\left(\frac{dU}{dx}P + D\frac{\partial P}{\partial x}\right) \equiv -\frac{\partial J}{\partial x}, \qquad (6.34)$$

reduces to a simple diffusion equation, namely, one has to solve equation (6.34) in each of three different regions of $U = const$, and then ensure the continuity of P and J on the boundaries of these regions. Continuity of probability current J, which according to equation (6.34) can be written as $J = -D\exp\left(-\frac{U}{D}\right)\frac{d}{dx}\left[\exp\left(\frac{U}{D}\right)P\right]$, means that at points x_i of the jumps of potentials, one gets

$$\exp\left[\frac{U\left(x_i+0\right)}{D}\right] P\left(x_i+0\right) = \exp\left[\frac{U\left(x_i-0\right)}{D}\right] P\left(x_i-0\right),$$

$$\frac{\partial P\left(x_i+0\right)}{\partial x} = \frac{\partial P\left(x_i-0\right)}{\partial x}.$$

(6.35)

The matching conditions (6.35) have to be complemented by reflected boundary conditions at the positions $\pm L$ of the walls, $\partial P\left(\pm L, t\right)/\partial x = 0$. It is easy to check [85] that the normalized stationary probability distribution function $P_{st}\left(x\right)$ in the absence of an external field, has the following form in each of three regions shown in Fig. 6.6,

$$P_{1,st} = \frac{\exp\left(U_1/D\right)}{2a + (L-a)\left[\exp\left(\frac{U_1}{D}\right) + \exp\left(\frac{U_2}{D}\right)\right]},$$

$$P_{2,st} = \frac{1}{2a + (L-a)\left[\exp\left(\frac{U_1}{D}\right) + \exp\left(\frac{U_2}{D}\right)\right]},$$

(6.36)

$$P_{3,st} = \frac{\exp\left(U_2/D\right)}{2a + (L-a)\left[\exp\left(\frac{U_1}{D}\right) + \exp\left(\frac{U_2}{D}\right)\right]}.$$

Since the two wells have equal width, the equilibrium population of the left (metastable) state is smaller than that of the right (stable) state by the ratio $\exp\left[(U_2 - U_1)/D\right]$.

An external force can be chosen as a periodic force acting on the well as is shown in Fig. 6.7, or as a random force acting on the barrier (Fig. 6.8). It turns out that for both cases, an external force attempts to equalize the populations at $t \to \infty$ (stabilizing the metastable state), even reversing the populations of these states. Our choice of the periodic signal does not introduce an additional force in equation (6.34). However, the periodic signal enters the matching conditions (6.35).

Starting from the periodic changes of the well (Fig. 6.7) [86], we assume that the amplitude of the external field A is small, $A < D$.

Fig. 6.7 Square double-well potential with an oscillating left well. [Reprinted figure with permission from Ref. 86. Copyright (1999) by the American Physical Society.]

Accordingly, we seek the solution of equation (6.34) in each region, $i = 1, 2, 3$, as

$$P_i = P_{i,st} + \sum_{l=1}^{\infty} \left(\frac{A}{D}\right)^l f_i^l(x, t) \tag{6.37}$$

where f_i^l is a periodic function of t which can be written in the following form

$$f_i^l = f_{i,0}^l + \sum_{k=1}^{\infty} \left\{ \left[f_{i,k}^l \exp\left(r_k x\right) + f_{i,k}^{-l} \exp\left(-r_k x\right) \right] + c.c. \right\} \exp\left(i\Omega k t\right),$$

$$r_k = \sqrt{\frac{i\Omega k}{D}}. \tag{6.38}$$

Fig. 6.8 Square double-well potential subject to the dichotomous fluctuations of the barrier height. [Reprinted figure with permission from Ref. 87. Copyright (2002) by the American Physical Society.]

Due to the boundary conditions on the walls the first order in time-independent corrections vanish, therefore, we keep also the non-oscillating second-order terms $\left(\frac{A}{D}\right)^2 g_i$ in (6.37). Then, one gets twelve parameters $(P_{i,st}, f_i, \tilde{f}_i\ g_i;\ i = 1, 2, 3)$ that satisfy twelve equations (two on the walls and two on each of two matching points for each order in the small parameter A/D). Omitting all calculations, we present here only the time-independent populations (up to second order in $(A/D)^2$) of the right well, $n_{r,\infty}$, and the left (oscillating) well, $n_{l,\infty}$,

$$
\begin{aligned}
n_{l,\infty} &= \int_{-1}^{0} P_1\left(x, t\right) dx = P_{1,st} + \left(\frac{A}{D}\right)^2 g_1 + \dots \\
n_{r,\infty} &= \int_{1}^{2} P_3\left(x, t\right) dx = P_{3,st} + \left(\frac{A}{D}\right)^2 g_3 + \dots
\end{aligned}
\tag{6.39}
$$

where $P_{1,st}$ and $P_{3,st}$ are defined in (6.36), and

$$g_1 = -\left[1 + \exp\left(\frac{U_2}{D}\right)\right]^{-1} ; \quad g_3 = -\left(P_{1,st} + P_{3,st}\right)\left[\frac{P_{1,st}}{4} + 2\operatorname{Re} f_1\right],$$

$$\operatorname{Re} f_1 = \frac{1 + \exp\left(-U_2/D\right)}{2\exp\left[\left(U_1 + U_2\right)/D\right]} \frac{4H^2 + \sin^2\left(\alpha_2\right)\sinh^2\left(\alpha_2\right)}{4H^2 - \left[\sinh^2\left(\alpha_2\right) - \sin^2\left(\alpha_2\right)\right]},$$

$$H = \frac{\left[1 + \exp\left(\frac{U_1}{D}\right) + \exp\left(\frac{U_2}{D}\right)\right]\left[\sinh^2\left(\alpha_1\right) + \cos^2\left(\alpha_1\right)\right]}{\exp\left(\frac{U_1}{D}\right) + \exp\left(\frac{U_2}{D}\right)}$$

$$+ \frac{\sinh^2\left(\alpha_2\right) - \sin^2\left(\alpha_2\right)}{4},$$

$$\alpha_n = \sqrt{\frac{n^2\Omega}{2D}}.$$

$$(6.40)$$

With the help of an external periodic field, one can increase the population of the left (metastable) state, or even reverse the populations. The latter, $n_{l,\infty} > n_{r,\infty}$, will occur when

$$P_{3,st} - P_{1,st} < \left(\frac{A}{D}\right)^2 \left(g_1 - g_3\right) \qquad (6.41)$$

which gives

$$\left(\frac{A}{D}\right)^2 \frac{1 + 2\exp\left(\frac{U_2}{D}\right)}{\exp\left(\frac{U_2}{D}\right) - \exp\left(\frac{U_1}{D}\right)} \left[-\frac{P_{1,st}}{4} - 2\operatorname{Re} f_1\right] > 1. \qquad (6.42)$$

The last inequality is obeyed when the factor in front of the brackets is large, and the expression in brackets is positive. The former occurs when $A > \sqrt{U_2 - U_1}$, and the latter holds for not too small frequencies.

Another form of stabilization of a metastable state is shown in Fig. 6.8, where the potential barrier undergoes dichotomous fluctuations of height Δ with flipping rate λ, i.e., the barrier for the right (left) well changes randomly between heights U_1, $U_1 + \Delta$ (U_2, $U_2 + \Delta$) [87]. The question is whether the dichotomous fluctuations of the potential barrier are able (like an external periodic field) to increase the stationary ($\frac{\partial P}{\partial t} = 0$) population of a metastable state.

Two probability functions $P^{\pm}(x)$, which define the probability to be at position x when the potential is $U_1 + \Delta$ or U_1, are related by the following equations

$$D\frac{d^2 P^{\pm}}{dx^2} + \lambda\left(P^{\mp} - P^{\pm}\right) = 0. \qquad (6.43)$$

The solutions of equations (6.43) have the following form

$$P^{\pm}_{1,2,3} = a_{1,2,3} + b_{1,2,3}\, x \pm c_{1,2,3} \sinh\left(\sqrt{\frac{2\lambda}{D}}x\right)$$

$$\pm d_{1,2,3} \cosh\left(\sqrt{\frac{2\lambda}{D}}x\right) \qquad (6.44)$$

where the subscripts $1, 2, 3$ relate to the three regions shown in Fig. 6.6. The coefficients in equation (6.44) have to be found from the matching conditions (6.35) and boundary conditions. Once these coefficients are found, one can calculate the populations $n_{1,\infty}$ of the metastable and $n_{3,\infty}$ of the stable state,

$$n_{1,\infty} = \int_{-L}^{-a} \left[P^+_1(x) + P^-_1(x)\right] dx; \quad n_{3,\infty}$$

$$= \int_a^L \left[P^+_3(x) + P^-_3(x)\right] dx \qquad (6.45)$$

Instead of writing the cumbersome formulas for $n_{1,\infty}/n_{3,\infty}$ [87], we shall bring their limiting forms for small and large amplitudes of noise, $\Delta/D \lessgtr 1$, namely, for $\frac{\Delta}{D} < 1$,

$$\frac{n_{1,\infty}}{n_{3,\infty}} = \exp\left(\frac{U_2 - U_1}{D}\right)$$

$$\times \left\{1 + \frac{2\left[\exp\left(-\frac{U_2}{D}\right) - \exp\left(-\frac{U_1}{D}\right)\right] \coth\left[\sqrt{\frac{2\lambda}{D}}(L-a)\right]}{1 + 16B_1B_2 + 4(B_1 + B_2)\coth\left[2\sqrt{\frac{2\lambda}{D}}a\right]}\left(\frac{\Delta}{D}\right)^2 + \dots\right\},$$

$$(6.46)$$

and for $\frac{\Delta}{D} > 1$,

$$\frac{n_{1,\infty}}{n_{3,\infty}} = \exp\left(\frac{U_2 - U_1}{D}\right)$$

$$\times \left\{ 1 + \frac{2\left[\exp\left(-\frac{U_2}{D}\right) - \exp\left(-\frac{U_1}{D}\right)\right]\coth\left[\sqrt{\frac{2\lambda}{D}}\,(L-a)\right]}{1 + 16B_1 B_2 + 4\left(B_1 + B_2\right)\coth\left[2\sqrt{\frac{2\lambda}{D}}\,a\right]} \left(\frac{\Delta}{D}\right)^2 + \ldots \right\},$$

$$(6.47)$$

where

$$B_1 = \exp\left(-\frac{U_1}{D}\right)\coth\left[\sqrt{\frac{2\lambda}{D}}\,(L-a)\right],$$

$$(6.48)$$

$$B_2 = \exp\left(-\frac{U_2}{D}\right)\coth\left[\sqrt{\frac{2\lambda}{D}}\,(L-a)\right].$$

Equation (6.47) shows that the ratio of populations in the left (metastable) and right (stable) states increases with increasing strength of the noise or the noise rate. For some values of parameters, the latter dependence is non-monotonic [87].

In conclusion, the population of a metastable state can be increased by the use of an external periodic field or by fluctuations of the barrier height.

Chapter 7

Overdamped double-well oscillator

Fluctuations of a double-well system driven by multiplicative $\xi(t)$ and additive $\eta(t)$ noise are described by the following equation

$$\frac{dx}{dt} = ax - bx^3 + x\xi(t) + \eta(t). \tag{7.1}$$

Equation (7.1) describes the motion of a particle in the potential $U(x) = -ax^2/2 + bx^4/4$ with two potential minima located at $x = \pm\sqrt{a/b}$ and a maximum at $x = 0$ (Fig. 7.1). The solution of the deterministic equation (7.1), with $\xi = \eta = 0$, is

$$x(t) = \frac{x(t=0)}{\sqrt{\frac{b}{a}x^2(t=0) + \left[1 - \frac{b}{a}x^2(t=0)\right]\exp(-2at)}} \tag{7.2}$$

which means that at $t \to \infty$ a particle will stay in one of the minima $x = \pm\sqrt{a/b}$, depending on the sign of $x(t=0)$. Our aim is to study equation (7.1) with noise included. Among many applications described by this equation are: statistical properties of the dye-laser laser light [88],[89], hydrodynamic instabilities [90], autocatalytic chemical reactions [91], current instabilities in parametrically excited circuits [92], and electrodynamic instabilities in nematic liquid crystals [74].

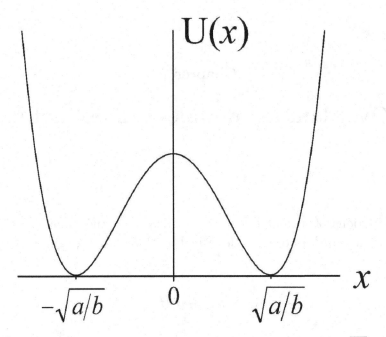

Fig. 7.1 A double-well potential with two minima located at $\pm\sqrt{a/b}$ and a maximum at $x = 0$.

7.1   Steady state

7.1.1   *White noise*

If the noise $\xi(t)$ and $\eta(t)$ in (7.1) are correlated white noise described by (6.4), the analysis of the stationary solutions of equation (7.1) is quite similar [93] to that presented in the previous section for the birth-death equation.

The stationary distribution function $P_{st}(x)$ for the Langevin equation (7.1) with $a = b = 1$, has the following form for $0 \le \kappa < 1$ [95]

$$P_{st}(x) = N\left(Dx^2 + 2\kappa\sqrt{D\alpha}x + \alpha\right)^{C-\frac{1}{2}}$$

$$\times \exp\left\{f(x) + \frac{E}{[(1-\kappa^2)D\alpha]^{\frac{1}{2}}} \arctan\frac{Dx + \kappa\sqrt{D\alpha}}{[(1-\kappa^2)D\alpha]^{\frac{1}{2}}}\right\}$$

$$(7.3)$$

where

$$f(x) = 2\kappa\sqrt{\frac{\alpha}{D}}\frac{bx}{D} - \frac{bx^2}{2D},$$

$$E = \kappa\sqrt{\frac{\alpha}{D}}\left[(4\kappa^2 - 1)\frac{b\alpha}{D} - a\right], \qquad (7.4)$$

$$C = \frac{b\alpha + aD - 4b\alpha\kappa^2}{2D^2}.$$

For perfect correlated noise, $\kappa = 1$,

$$P_{st}(x) = N\left(\sqrt{\alpha} + \sqrt{D}x\right)^{2C_{\kappa=1}-1} \exp\left[f(x)_{\kappa=1} + \frac{E_{\kappa=1}}{Dx + \sqrt{D\alpha}}\right].$$

$$(7.5)$$

The extrema of $P_{st}(x)$ obey the following equation

$$bx^3 + (D - a)x + \kappa\sqrt{D\alpha} = 0, \quad \text{for } 0 \le \kappa < 1. \qquad (7.6)$$

For non-correlated noise, $\kappa = 0$, equation (7.6) has two roots, $x = 0$ and $x = \sqrt{(a - D)/b}$ [94]. For small intensity of noise, $D < a$, $P_{st}(x)$ has a single maximum at $x = \sqrt{(a - D)/b}$ which is shifted to smaller values of x when D increases. For $D > a$, the maximum of $P_{st}(x)$ disappears and P_{st} goes to infinity at $x \to 0$..

Some interesting results follow from the analysis of equations (7.3)-(7.6):

1. The presence of correlation between noise, $\kappa \ne 0$, leads to essential changes in the steady state distribution, namely, the distribution function $P_{st}(x)$ which had a symmetric bimodal structure at $\kappa = 0$, becomes non-symmetric for $\kappa \ne 0$.

2. As is well known [74], the extrema of $P_{st}(x)$ are shifted by multiplicative noise but not by additive noise. However, as one can see from (7.6), in the case of the correlation between these two sources of noise, the extrema of $P_{st}(x)$ are affected not only by multiplicative noise but also by additive noise.

3. Without noise, the steady state distribution has two minima and one maximum ($D = \alpha = 0$ in (7.6) gives $x = 0, \pm\sqrt{a/b}$), while the presence of noise may change this distribution from bimodal

to unimodal. The latter is influenced by both multiplicative and additive noise and takes place when the three solutions of equation (7.6) reduce to one solution, i.e., when

$$\frac{\kappa^2 D\alpha}{4} + \frac{(D-a)^3}{27b} = 0. \tag{7.7}$$

4. For the case of perfectly correlated noise, $\kappa = 1$, the steady state distribution (7.5) exhibits a divergence at $x = -\sqrt{\alpha/D}$, and the function $P_{st}(x)$ exhibits a different shape of divergence for $b\alpha/aD > 1$ and $b\alpha/aD < 1$.

5. The first moment $\langle x \rangle_{st}$ decreases with increasing κ. The effect of κ on $\langle x \rangle_{st}$ is more pronounced for $D \geq \alpha$ than for the case of $D < \alpha$.

A slightly more complicated case, where the delta-correlation between noise (6.4) is replaced by the exponential form correlation with the correlation time as an additional parameter, has been considered in [95], [96].

7.1.2 *Dichotomous noise*

For the Langevin equation with multiplicative noise

$$\frac{dx}{dt} = ax - bx^3 + \xi(t)x \tag{7.8}$$

one can get an explicit formal solution by the change of variables indicated above in (3.17)-(3.18), namely, in terms of the new variable $y = x^{-2}$, equation (7.8) takes the following form

$$\frac{dy}{dt} = 2b - 2\left[a + \xi(t)\right]y. \tag{7.9}$$

This linear equation can be solved exactly which allows one to find the dynamic (time-dependent) moments $\langle y^n(t) \rangle$, i.e. the inverse moments of $x(t)$. Although it does not help to solve the original equation (7.8), this approach serves [97] as a useful test for the validity of an approximate solution of (7.8) for small values of the correlation time.

Let us restrict ourselves to multiplicative dichotomous noise in (7.8). Analysis similar to (6.16)-(6.21) can be performed for the Langevin equation (7.8), which gives the stationary distribution function

$$P_{st}(x) = N \, x^{-1+\frac{\lambda a}{(\sigma^2-a^2)}} \, \left|a-\sigma-bx^2\right|^{-1+\frac{\lambda}{2(a-\sigma)}} \, \left|a+\sigma-bx^2\right|^{-1+\frac{\lambda}{2(a+\sigma)}}$$

(7.10)

and the extrema of (7.10)

$$a - bx^2 - \frac{\sigma^2}{\lambda} + \frac{2}{\lambda}\left(a - bx^2\right)\left(a - 3bx^2\right) - \frac{1}{\lambda}\left(ax - bx^3\right)^2 = 0. \quad (7.11)$$

Linear stability analysis and the calculation of equilibrium moments have been performed for this case in [98].

7.2 Eigenfunction expansion of the Fokker-Planck equation

In addition to the stationary solution of the Fokker-Planck equation, one can find the full solution of this equation by introducing the following ansatz

$$P(x,t) = P(x)\exp(-\epsilon t) \tag{7.12}$$

which changes the type of the Fokker-Planck differential equation from a partial to an ordinary differential equation [91].

Taking into account (7.12), the time-dependent Fokker-Planck equation which corresponds to the Langevin equation (7.8) has the following form

$$\frac{d^2}{dx^2}\left[x^2 P(x)\right] - \frac{d}{dx}\left\{\left[\left(1 + \frac{a}{D}\right)x - \frac{bx^3}{D}\right]P(x)\right\} = -\frac{\epsilon}{D}P(x).$$

(7.13)

Solutions of this equation have a different form for $\epsilon \leq \frac{a^2}{4D}$ and $\epsilon > \frac{a^2}{4D}$ [91]. In the former case, the eigenfunctions of equation (7.13) are

$$P_n(x) = x^{-1+\frac{a}{2D}-2n} \exp\left(-\frac{bx^2}{2D}\right) {}_1F_1\left(-n, 2v_n, \frac{bx^2}{2D}\right) \qquad (7.14)$$

where ${}_1F_1$ is the confluent hypergeometric function [99],

$$2v_n = 1 + \frac{a}{2D} - 2n > 0 \qquad (7.15)$$

and the eigenvalues spectrum

$$\epsilon_n = 4nD\left(\frac{a}{2D} - n\right). \qquad (7.16)$$

The distribution function (7.14) has a singularity at $x = 0$ for $-1 + \frac{a}{D} - 2n < 0$ which disappears if one includes additive noise in the original Langevin equation.

For $\epsilon > \frac{a^2}{4D}$, the eigenfunctions are

$$P_n(x) = x^{-1+\frac{a}{2D}} \exp\left(-\frac{bx^2}{2D}\right) [f(\mu) + f(-\mu)] \qquad (7.17)$$

with

$$f(\mu) = \frac{\Gamma(-2\mu)}{\Gamma\left(\frac{1}{2} - \mu - v\right)} x^{2\mu} {}_1F_1\left(\mu - v + \frac{1}{2}, 2\mu + 1, \frac{bx^2}{2D}\right),$$

$$\qquad (7.18)$$

$$\mu = \frac{i}{\sqrt{4D}}\sqrt{\epsilon - \frac{a^2}{4D}}; \quad v = \frac{1}{2}\left(\frac{a}{2D} + 1\right)$$

and eigenvalues cover the continuous part of the spectrum for all real values of ϵ satisfying the condition $\epsilon > \frac{a^2}{4D}$.

Note the remarkable property of the eigenvalue spectrum which consists of both discrete and continuum branches.

7.3   Matrix continued fraction method

In the previous section, we considered the general method of solution the Fokker-Planck equation. However, in many cases a system is subjected to a periodic force, and it is then worthwhile to use matrix continued fractions. The typical equation of motion with white noise $\xi(t)$ has the following form

$$\frac{dx}{dt} = f(x) + A\sin(\Omega t) + \xi(t) \tag{7.19}$$

where $f(x)$ is the arbitrary function of x. The Fokker-Planck equation corresponding to (7.19) is

$$\frac{\partial P(x,t)}{\partial t} = -\frac{\partial}{\partial x}\{[f(x) + A\sin(\Omega t)]P(x,t)\} + D\frac{\partial^2 P(x,t)}{\partial x^2}. \tag{7.20}$$

As was shown in chapter 5.2 for the linear function $f(x) = ax$, these equations possess exact solutions for the first two moments. In the general case, one can use the periodicity in time of the coefficients in the differential equation (7.20), which means [100] that (7.20) possesses Floquet-type solutions

$$P(x,t) = \exp(\alpha t)p_\alpha(x,t) \tag{7.21}$$

where the functions $p_\alpha(x,t)$ are periodic in time, i.e., $p_\alpha(x,t+T) = p_\alpha(x,t)$, where $T = 2\pi/\Omega$. Expanding the periodic function $p_\alpha(x,t)$ in a Fourier series

$$p_\alpha(x,t) = \sum_{n=-\infty}^{n=\infty} c_n^\alpha(x)\exp(i\Omega t), \tag{7.22}$$

and substituting (7.21) and (7.22) into (7.20), one gets

$$\left[-\frac{df}{dx} - f\frac{d}{dx} + D\frac{d^2}{dx^2} - in\Omega - \alpha\right]c_n^\alpha(x)$$
$$-\frac{iA}{2}\left[c_{n+1}^\alpha(x) - c_{n-1}^\alpha(x)\right] = 0. \tag{7.23}$$

It is convenient [101] to expand the coefficients $c_n^\alpha(x)$ into the complete orthogonal set of eigenfunctions

$$c_n^\alpha(x) = \sum_{m=0}^{\infty} c_n^m \psi_m(x). \tag{7.24}$$

As an example, for the double-well potential $f(x) = x - x^3$, it is convenient to use the Hermite polynomials $H_n(\rho x)$,

$$\psi_m(x) = \sqrt{\frac{\rho}{n! 2^n \sqrt{\pi}}} \exp\left(-\frac{\rho^2 x^2}{2}\right) H_n(\rho x) \qquad (7.25)$$

where ρ is an arbitrary parameter. Inserting (7.24) and (7.25) into (7.23) and using the orthogonality and recursion properties of the Hermite polynomials, one gets

$$\sum_{\overline{m}=0}^{\infty} \left[Q_{m,\overline{m}}(n) c_n^{\overline{m}} + Q_{m,\overline{m}}^+ c_{n+1}^m + Q_{m,\overline{m}}^- c_{n-1}^m \right] = 0 \qquad (7.26)$$

with the complex matrices

$$Q_{m,\overline{m}}(n) = \left[-\frac{df}{dx} - \frac{d}{dx} + D\frac{d^2}{dx^2} \right]_{m,\overline{m}} - in\Omega\delta_{m,\overline{m}} - \alpha\delta_{m,\overline{m}},$$

$$Q_{m,\overline{m}}^+(n) = -Q_{m,\overline{m}}^-(n) = -\frac{iA}{2} B_{m,\overline{m}}. \qquad (7.27)$$

The cumbersome expressions for the matrices

$$\left[-\frac{df}{dx} - f\frac{d}{dx} + D\frac{d^2}{dx^2} \right]_{m,\overline{m}} \qquad (7.28)$$

and $B_{m,\overline{m}}$ are given in [101].

Hence the problem is reduced to analysis of the tridiagonal recursion relations

$$Q_n(\alpha)\overline{c_n} + Q_n^+ \overline{c_{n+1}} + Q_n^- \overline{c_{n-1}} = 0 \qquad (7.29)$$

which appear in different problems [102]. The solution of (7.29) can be found by introducing the rates

$$S_n = \frac{\overline{c_{n+1}}}{\overline{c_n}}. \qquad (7.30)$$

Inserting (7.30) into (7.29) leads to

$$\overline{c_n} = -\frac{Q_n^-}{Q_n + Q_n^+ S_n} \overline{c_{n-1}} \equiv S_{n-1}\overline{c_{n-1}}. \qquad (7.31)$$

Comparing (7.30) and (7.31) yields a recurrent relation for S_n,

$$S_n = -\frac{Q_{n+1}^-}{Q_{n+1} + Q_{n+1}^+ S_{n+1}} \tag{7.32}$$

which, starting from $S_0 = c_1/c_0$, can be used again and again. Finally, one gets

$$c_n = S_{n-1} S_{n-2} ... S_0 c_0. \tag{7.33}$$

Another method for solving equation (7.29) is the transition from a difference to a differential equation by replacing the integers n by the continuous variable and using a Taylor expansion

$$\overline{c_{n+1}} - \overline{c_{n-1}} \cong 2\frac{d\overline{c_n}}{dn} + O\left(\frac{2}{3}\frac{d^3\overline{c_n}}{dn^3}\right). \tag{7.34}$$

For some special form of the matrices $Q(\alpha)$ and Q^+, this method gives quite accurate results [103].

7.4 Mean first-passage time

A particle located in, say, the left minimal position, $x_1 = -\sqrt{a/b}$, of the potential shown in Fig. 7.1 will stay there forever unless some external force, such as an external noise, causes the particle to leave this minimum and to jump to the right minimal position, $x_2 = \sqrt{a/b}$. For analysis of this and similar processes, let us define the mean first-passage time T in terms of probability that the random variable x starting from the value x_0 reaches the value x_1 for the first time in a time interval between t and $t + dt$. "For the first time" is taken to mean that after reaching the value x_1, the process is finished, a particle is no longer counted, i.e. the point x_1 is the absorbing point with vanishing probability $P(x_1, T) = 0$. The latter condition plays a role of the boundary condition for the Fokker-Planck equation.

It is not surprising, therefore, that the mean first-passage time T can be expressed [28] in terms of the solution of the Fokker-Planck equation corresponding to (7.1).

The probability that the first-passage time T from $x = x_1$ to $x = x_2$ is greater than t is defined by the so-called survival probability

$S(t)$,

$$S(t) = \int_{x_1}^{x_2} P(x,t)\, dx. \tag{7.35}$$

All the statistics of the first-passage time can be computed from $S(t)$. For example, the mean free-passage time is defined as

$$T = \int_0^\infty S(t)\, dt. \tag{7.36}$$

In the case of two correlated sources of white noise (6.4), T is given by [104]

$$T = \int_{x_1}^{x_2} dx \left[\left(Dx^2 + 2\kappa\sqrt{D\alpha}x + \alpha \right) P_{st}(x) \right]^{-1} \int_{-\infty}^{x} dy\, P_{st}(y). \tag{7.37}$$

For noise with strengths D and α smaller than the potential barrier in Fig. 7.1, the mean free-passage time becomes independent of the initial condition $x\,(t=0)$ and of the final point x_2 (it is enough to reach the point $x=0$ of potential maximum in order to reach the point x_2). The explicit expression for T [105] for $0 \le \kappa < 1$, has the following form

$$T = \sqrt{2\pi} \left| \frac{aD}{b\alpha} - 2\kappa\sqrt{\frac{aD}{b\alpha}} + 1 \right|^{-C} \exp\left[-\frac{a}{2D} + \frac{2b\kappa\sqrt{D\alpha}}{D^2} \right.$$

$$\left. + \frac{E}{D}\left(\arctan\frac{\kappa}{\sqrt{1-\kappa^2}} - \arctan\frac{\kappa - \sqrt{aD/b\alpha}}{\sqrt{1-\kappa^2}} \right) \right] \tag{7.38}$$

and for $\kappa = 1$

$$T = \sqrt{2\pi} \left| 1 - \sqrt{\frac{aD}{b\alpha}} \right|^{-2C_{\kappa=1}} \exp\left[-\frac{a}{2D} + \frac{2b\sqrt{D\alpha}}{D^2} + \frac{E_{\kappa=1}}{D\sqrt{\alpha} - \alpha\sqrt{D}} \right] \tag{7.39}$$

where C and E are defined in (7.4).

Analysis of formulas (7.38)-(7.39) shows that [105]

1. For fixed $\alpha\,(D)$, T increases with increasing κ, but decreases when $D\,(\alpha)$ increases. The effect of κ on T is in good agreement

with the transition from a bimodal to unimodal structure of $P_{st}(x)$ as κ increases, so that the effect of κ is to make T decrease.

2. The mean first-passage time is biggest for $D = \alpha$, as compared with $D \gtrless \alpha$.

3. For the perfectly correlated noise $\kappa = 1$, the dependence of T on D and on α is quite different. When α is fixed, T increases with increasing D for $\alpha > D$, and decreases for $\alpha < D$. On the other hand, when D is fixed, T increases with increasing α for $\alpha < D$, and decreases for $\alpha > D$.

If the correlation time τ of the correlation between additive and multiplicative noise is nonzero, one can study the dependence of the mean first-passage time T on τ [106], [107].

7.5 Response to a periodic force (stochastic resonance)

In section 5.3, we considered stochastic resonance in linear systems. However, this phenomenon usually appears under the heading of "nonlinear phenomena", describing a nonlinear (usually bistable) system driven by a combination of an additive random and periodic forces

$$\frac{dx}{dt} = ax - bx^3 + \eta(t) + A\cos(\Omega t). \qquad (7.40)$$

The amplitude of the periodic force A in (7.40) is assumed to be sufficiently small that in the absence of noise this force is unable to move a particle from one well to the other. This condition is satisfied when A is smaller than the barrier height $\Delta U = U(0) - U\left(\sqrt{\frac{a}{b}}\right) = \frac{a^2}{4b}$, i.e., $A < \frac{a^2}{4b}$. Although this signal is unable by itself to transfer a particle through the barrier, it makes it easier for the random signal to induce a transfer. The latter occurs when the deterministic frequency Ω approaches a characteristic frequency of the system without the periodic force. The appropriate frequency to characterize the equation (7.40) in the absence of a periodic force is the Kramers rate $r = r_0 \exp(-\Delta U/D)$, where r_0 has the dimensions of frequency. Most curious is the role of noise in this process. Too small noise cannot help to overcome the threshold (barrier height), while too strong noise destroys the signal. Therefore, an intermediate strength of the noise will be the most useful.

The manifestation of stochastic resonance can roughly be divided into two classes. The first relates to $x(t)$ at the characteristic frequency and can be expressed in terms of different functions that are calculated in terms of $x(t)$. The mathematical form of this relation is usually represented in terms of Fourier components of these quantities which behave non-monotonically when evaluated as a function of the noise strength D. A second group of physically interesting parameters which can show resonant behavior are characteristic times, such as the reciprocal of the switching time between the wells in the model defined by equation (7.40) [108]. This can also depend non-monotonically on the strength D of the noise. Still another measure of stochastic resonance is the input energy, i.e., the work done by an external field, which turns out to be non-monotonic function of both noise strength [109] and frequency of an external field [110].

In principle, one can use stochastic resonance for the resolution of the long-standing engineering problem of the detection and amplification of weak signals embedded within a large noise background. Thereby, the weak signal increases at the expense of noise. Since noise is found everywhere in nature, the question arises if "weak" signals are actually weak[1].

There are several comprehensive reviews on stochastic resonance [111], [112], [113], [114], containing the full analysis of equation (7.40). Therefore, we have restricted ourselves to the case of two sources of noise,

$$\frac{dx}{dt} = ax - bx^3 + \xi(t)x + \eta(t) + A\cos(\Omega t). \qquad (7.41)$$

Let us start with the case of non-correlated white noise of strength D and α with $\kappa = 0$ in (6.4) [115]. Equation (7.41) differs from (7.1) by an additional periodic force $A\cos(\Omega t)$. We assume, as it is usually done in the theory of stochastic resonance, that both amplitude A

---

[1]Pursuing these questions, one can think of far-reaching applications, such as the real danger of the "weak" radiation from our computers and cellular telephones, the ability of telepathic transmission of "weak" signals, etc.

and frequency Ω are small enough. The smallness of amplitude means that in the absence of any noise, the periodic force is unable to force a particle to move from one well to the another, while the smallness of the frequency (the so-called adiabatic limit) allows the system to reach local equilibrium in the period of $2\pi/\Omega$. In the adiabatic limit, one can simply add the slowly changing periodic term as a constant into an expression obtained for the static potential, and get the quasi-steady-state distribution function $P_{st}(x,t)$,

$$P_{st}(x,t) = N \left| Dx^2 + \alpha \right|^{\frac{a}{2D} + \frac{b\alpha}{2D^2} - \frac{1}{2}}$$
$$\times \exp\left[-\frac{bx^2}{2D} + \frac{A}{\sqrt{D\alpha}} \arctan\left(\sqrt{\frac{D}{\alpha}}x \right) \cos(\Omega t) \right]. \quad (7.42)$$

The distribution function $P_{st}(x,t)$, according to (7.37), allows one to find the first-passage time T in terms of the first-passage time T_0 in the absence of a periodic signal

$$T = T_0 \exp\left[A\sqrt{\frac{D}{\alpha}} \arctan\left(\sqrt{\frac{aD}{b\alpha}} \right) \cos(\Omega t) \right] \quad (7.43)$$

where

$$T_0 = \frac{\sqrt{2}\pi}{a} \left| 1 + \frac{aD}{b\alpha} \right|^{\frac{a}{2D} + \frac{b\alpha}{2D^2}} \exp\left(-\frac{a}{2D} \right). \quad (7.44)$$

Equations (7.42) and (7.44) reduce to (7.3) and (7.38) with non-correlated noises, $\kappa = 0$.

Stochastic resonance is characterized by the so-called signal-to-noise ratio which is defined in (5.28). For equation (7.41), one gets [115]

$$R = \frac{\pi A^2}{4T_0 D\alpha} \left(\arctan\sqrt{\frac{aD}{b\alpha}} \right)^2 \left[1 - \frac{A^2 \left(\arctan\sqrt{\frac{aD}{b\alpha}} \right)^2}{2D\alpha(\Omega^2 T_0^2 + 1)} \right]^{-1}. \quad (7.45)$$

Analysis of (7.45) shows that effects of the multiplicative and additive noise strengths on the signal-to-noise ratio R are opposed to each other, namely, the peak of R increases (decreases) with increase of D (α) when $\alpha(D)$ is maintained fixed. In both cases, the peak increases with increasing amplitude A of the periodic signal. However, for the fixed α (and not for the fixed D), the second maximum in R can appear when the amplitude of the periodic signal A is increased, replacing thereby a "single stochastic resonance" (SSR) by a "double stochastic resonance" (DSR). This phenomenon was found for the first time by the analysis of traditional stochastic resonance equation (7.40) for sufficient low frequency Ω and for the increasing amplitude A of the input signal [113].

These results have been obtained for non-correlated noise. One can include white noise-like correlations using equations (6.4) with $\kappa \neq 0$. It turns out [115], that for $\kappa \neq 0$, the signal-to-noise ratio R depends not only on D, α and κ, but also on the initial condition $x(t=0)$ of the system. Moreover, for full correlations $\kappa = -1$ and $x(t=0) = \sqrt{a/b}$ (or $\kappa = 1$ and $x(t=0) = -\sqrt{a/b}$), the peak becomes very narrow and large $(R \approx 10^4)$. On the other hand, for $\kappa = -1$ and $x(t=0) = -\sqrt{a/b}$ (or $\kappa = 1$ and $x(t=0) = \sqrt{a/b}$), the peak is broadened and becomes very small $(R \approx 10^{-7})$.

Another way to complicate the calculations (7.42)-(7.45) is to replace the white multiplicative noise (6.4) by the colored Ornstein-Uhlenbeck noise,

$$\langle \xi(t_1) \xi(t_2) \rangle = \frac{Q}{\tau_1} \exp\left(-\frac{|t_1 - t_2|}{\tau_1}\right),$$

$$\langle \eta(t_1) \eta(t_2) \rangle = 2\alpha\delta(t_2 - t_1). \tag{7.46}$$

We omit the cumbersome formulas for $P_{st}(x,t)$, T and R obtained in [116], restricting our consideration to the qualitative results. For $\tau_1 = 0$ (white noise) or small τ_1, the signal-to-noise ratio R has one peak as a function of the noise strength ratio Q/α, which increases as τ_1 is increased. In addition to this SSR, the second peak appears at a smaller value of Q/α showing the DSR. When $Q/\alpha \to 0$, R approaches a finite value, and $R \to 0$ when $Q/\alpha \to \infty$. The joint action of κ (correlation between noise) and τ_1 (correlation rate of multiplicative noise) on the signal-to-noise ratio has been considered in [116].

In addition to noise of types (6.4) and (7.46), the more complicate case of white additive noise (7.46) and color multiplicative noise with color coupling between noises

$$\langle \xi(t_1)\xi(t_2)\rangle = \frac{Q}{\tau_1}\exp\left(-\frac{|t_1-t_2|}{\tau_1}\right),$$

$$\langle \xi(t_1)\eta(t_2)\rangle = \frac{\kappa\sqrt{\alpha Q}}{\tau_2}\exp\left(-\frac{|t_1-t_2|}{\tau_2}\right)$$

(7.47)

has been considered in [117], where the authors studied the dependence of the signal-to-noise ratio R on each of the three parameters (the multiplicative noise correlation time τ_1, the correlation time τ_2 of the coupling between noise and the coupling parameter κ), keeping the other two parameters fixed. It turned out that for all initial conditions, when the value of τ_1 and κ are increased, there appears a transition from one peak to two peaks, and then back to one peak. However, when the value of τ_2 is increased, there appears, depending on initial conditions, a transition either from one peak to two or from two peaks to one, but not two successive transitions. The DSR phenomenon (appearance of two peaks in signal-to-noise ratio as a function of noise strength) can also be induced by dichotomous noise in bistable systems (7.40) [118].

All the preceding examples were related to the simultaneous action of additive and multiplicative noise on a particle moving in a symmetric potential barrier ($U = -\frac{ax^2}{2} + \frac{bx^4}{4}$ in equation (7.41)). The question arises regarding the form of the stochastic resonance for an asymmetric potential. Two special forms of asymmetric potentials have been considered in [119]. The first is of the form $U = -\frac{x^2}{2} - \frac{ax^3}{3} + \frac{x^4}{4}$, replacing a symmetric potential $U = -\frac{ax^2}{2} + \frac{bx^4}{4}$ in the traditional equation (7.40), which leads to equation

$$\frac{dx}{dt} = x + ax^2 - x^3 + \eta(t) + A\cos(\Omega t);$$

(7.48)

the second is of the form $U = \frac{x^4}{4} + \frac{2x^3}{3}$ in the presence of both additive and multiplicative noise, replacing equation (7.41) by

$$\frac{dx}{dt} = -2x^2 - x^3 + \xi(t)x + \eta(t) + A\cos(\Omega t).$$

(7.49)

Analysis of equation (7.48) shows that the peak of the signal-to-noise ratio as a function of noise strength becomes smaller with increase of a, i.e., the asymmetry of the bistable potential can weaken the phenomenon of stochastic resonance. As for equation (7.49), it turns out that stochastic resonance occurs not for all values of the parameters in this equation. This phenomenon is absent when there is no multiplicative noise. It is worth mentioning that these results are sensitive to special forms of the asymmetric potential.

An alternative form of an asymmetric potential of the form

$$U\left(x\right) = \frac{ax^2}{2} + \frac{bx^4}{4} - hx \tag{7.50}$$

has been considered in [120] which leads to the equation

$$\frac{dx}{dt} = -ax - bx^3 + h + \xi\left(t\right)x + \eta\left(t\right) \tag{7.51}$$

where $\xi\left(t\right)$ and $\eta\left(t\right)$ are white noise of strengths D and α, respectively. It turns out [120] that multiplicative noise causes an exponential enhancement of the escape time out of a metastable state. The symmetry breaking field $h \neq 0$ decreases the escape time, although this influence becomes less important for large D and α.

7.6   Fluctuating potential barrier (resonance activation)

In section 6.2, we considered the stationary properties of metastable states subject to fluctuations of the potential barriers which leads to noise-enhanced stability. Here, we consider similar phenomena for a bistable potential which will result in a minimum of the mean escape time. The minimal value of this time is of the order of the inverse fluctuation rate of the barrier, suggesting a resonant character of this phenomenon which is called resonant activation. The distinction between resonance activation and the stochastic resonance considered in the previous section is that the modulation of a potential barrier is random in the former case but deterministic in the latter case.

7.6.1 *Piece-wise linear potential*

Just as a simplified piecewise squared potential has been used in the previous chapter to obtain the exact analytical solution, one can use a piecewise linear potential for the description of the resonance activation. This potential is shown in Fig. 7.2 under the assumption that the potential increases to infinity at $x = \pm L$ and fluctuates randomly between two configurations $V_{\pm}(x)$ with flipping rate λ (dichotomous noise). Such a model has been analyzed in [60].

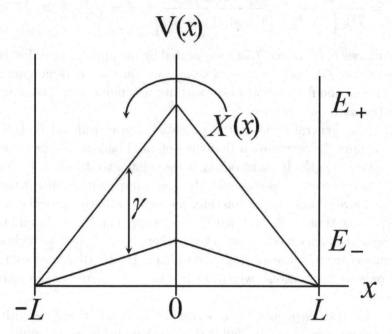

Fig. 7.2 The triangle piece-wise potential subject to the dichotomous fluctuations of the barrier height. [Reprinted figure with permission from Ref. 60. Copyright (1992) by the American Physical Society.]

The mean first-passage time T is considered for a particle subject to white noise, starting at the bottom of a potential well at $x = L$ in Fig. 7.2, and reaching the top of the barrier at $x = 0$. The boundary conditions are assumed to be reflecting at $x = L$ and absorbing at $x = -L$. Using equations (7.35)-(7.36), for the case where the midpoint of the barriers fluctuates between $\pm E$, one obtains in terms of the dimensionless variables $\tau = \frac{TD}{L^2}$ and $\varepsilon = \frac{\lambda L^2}{D}$ [60]

$$\tau = A_- \left[-\frac{E}{Dk} \left(1 - \exp(k) \right) + \frac{\varepsilon D k}{E} - \frac{E}{D} \right]$$

$$+ A_+ \left[\frac{E}{Dk} \left(1 - \exp(-k) \right) - \frac{\varepsilon D k}{E} - \frac{E}{D} \right], \qquad (7.52)$$

where

$$A_\pm = \frac{2\varepsilon \left[\pm \exp \left(\pm k \right) - k \right] \pm \frac{E^2}{D^2}}{\frac{2Ek}{D} \left\{ 1 + \frac{2\varepsilon D^2}{E^2} \left[2\varepsilon \cosh \left(k \right) + \frac{E^2}{D^2} \right] \right\}}, \quad k = \sqrt{\frac{E^2}{D^2} + 2\varepsilon}. \quad (7.53)$$

Analysis of equation (7.52), supported by the Monte Carlo simulations, shows [60] that the mean free-passage time is a non-monotonic function of the fluctuation rate, reaching minimum at a "resonant" rate (resonant activation).

A quite general form of the potential was considered in [121], making only the assumption that the potential barrier is higher than the noise strength. It turns out that resonant activation is a typical phenomenon which has a simple physical interpretation. For a very fast and a very slow fluctuation rate, the escape rate (reciprocal mean free-passage time) is determined by the average barrier height and by the highest barrier height, respectively. For the intermediate regime, the rate is given by the average rate which is greater than the rate for fast or slow fluctuations, which assures the maximum of the escape rate.

An interesting phenomenon occurs when, in addition to additive white noise considered in [60],[121], another dichotomous additive noise acts on the particle crossing a fluctuating barrier. It turns out [122] that the mean free-passage time displays two resonant activations : one (as described above) as a function of the flipping rate of the fluctuating potential barrier, and the other as a function of the correlation rate of dichotomous noise. Dichotomous noise can weaken the former resonant activation but enhances the latter one.

An additional complication of the original model of resonant activation [60] involves the introduction of multiplicative noise that supplements the additive noise and is correlated with it [123]. It turns out that additive and multiplicative noises can weaken the resonant activation while the correlation between them can enhance it. The influence of multiplicative noise on the resonant activation is much stronger than that of additive noise.

7.6.2 *Phenomenological model*

The general feature of stochastic activation is supported not only by the complication of the original model [60] made in [122], but also by its simplification. Resonant activation has been obtained [124] for a particle randomly switched at a rate λ between two states called $+$ and $-$, respectively. In addition, a particle may leave each of these states at time t with probabilities $\psi_+(t)$ and $\psi_-(t)$. Three different forms of these functions have been considered in [124].

a) The particles switch between two states with characteristic rates k_+ and k_-. For this case, the mean free-passage time is equal to

$$T = \frac{(k_+ + k_-)/2 + 2\lambda}{k_+ k_- + \lambda(k_+ + k_-)} \qquad (7.54)$$

which is a monotonic function of λ decreasing from $\frac{1}{2}(1/k_+ + 1/k_-)$ for $\lambda = 0$ to $2/(k_+ + k_-)$ for $\lambda = \infty$, i.e., there is no resonant activation.

b) The probability densities for leaving the states are $\psi_+(t) = \delta(t - k_+^{-1})$ and $\psi_-(t) = \delta(t - k_-^{-1})$. For this case,

$$T = \frac{1 + 2\exp[\lambda(1/k_+ + 1/k_-)] - 3/2[\exp(\lambda/k_+) + \exp(\lambda/k_-)]}{\lambda[\exp(\lambda/k_+) + \exp(\gamma/k_-) - 1]}.$$
$$(7.55)$$

It follows from equation (7.55) that for $\lambda \to 0$, $T \to (k_+^{-1} + k_-^{-1})/2$, and that T diverges for $\lambda \to \infty$. However, T goes through a minimum when k_+ and k_- differ by at least 3.75, showing thereby the resonant activation.

c) For the case of $\psi_+(t) = \psi_-(t)$, resonant activation exists for all forms of this function.

7.6.3   *Coherent stochastic resonance*

In addition to the simple model described in the previous section, there is another simple linear system which behaves similarly regarding resonant activation. This is one-dimensional diffusion on a segment terminated by one or two traps with a periodic force of small amplitude so that the particle cannot reach a trap in the absence of noise [125]. Although the system remains linear, the trapping boundaries produce additional states and the particle can be trapped or untrapped. The mean free-passage time to trapping has been shown to vary non-monotonically as a function of both the frequency [125] and the amplitude [126] of the periodic force (coherent stochastic resonance). An external periodic force supplies the characteristic frequency just like the flipping rate between two positions of the potential barriers in the case of resonant activation. Still another possibility for introducing the characteristic frequency was considered in [127], where a random walk on a finite interval terminated by one or two traps occurred in such a way that at any time, each site has one or two transition rates changing at random. Such system also shows coherent stochastic resonance. The explanation of this phenomenon is similar to that for resonant activation, and connected with the different behavior of the mean free-passage time at low and high frequencies of a periodic field [128]. At very low frequency, $\sin(\Omega t) \approx \Omega t$, and when the frequency increases from zero, T necessarily decreases, becoming smaller than its value at $\Omega = 0$. On the other hand, for $\Omega \to \infty$, the rate of oscillation increases, the particle is less and less influenced by the periodic force, and T approaches its value of a pure diffusive process as it was the case for $\Omega = 0$. It is clear, therefore, that T exhibits a minimum at some frequency. If a periodic force is replaced by a periodic [128] or random [129] telegraph signal, the different behavior at small frequency eliminates coherent stochastic resonance.

Chapter 8

Harmonic oscillator with additive noise

8.1 Internal and external noise

The equation of motion for $x(t)$ has a different form for external and internal noise $\xi(t)$. For external noise, the friction coefficient γ and the correlation function of the noise are independent, and the equation of motion has the following form

$$\frac{d^2x}{dt^2} + 2\gamma\frac{dx}{dt} + \omega^2 x = \xi(t) \tag{8.1}$$

while for internal noise which usually scales with the inverse system size, the dissipation and noise stem from the same source, and the system will finally reach an equilibrium state. Then, equation (8.1) has to be rewritten as

$$\frac{d^2x}{dt^2} + 2\int_0^t \gamma(t-t_1)\frac{dx}{dt}(t_1)\,dt_1 + \omega^2 x = \xi(t) \tag{8.2}$$

where, due to the fluctuation-dissipation theorem,

$$\langle \xi(t)\xi(t+\tau)\rangle = 4\kappa T\gamma(\tau). \tag{8.3}$$

8.2   White and dichotomous noise

After averaging equation (8.1), one immediately finds for the first moment

$$\langle x\,(t) \rangle = \exp\,(-\gamma t)$$

$$\times \left[\frac{\gamma x\,(t=0) + \frac{dx}{dt}(t=0)}{\omega_1} \sin\,(\omega_1 t) + x\,(t=0)\cos\,(\omega_1 t) \right],$$

$$\omega_1 = \sqrt{\omega^2 - \gamma^2}.$$

$$(8.4)$$

The stationary distribution function $P_{st}\,(x)$ can easily be found from the Fokker-Planck equation. For the equilibrium system (8.2), $P_{st}\,(x)$ has the canonical equilibrium distribution,

$$P_{st}\left(x, \frac{dx}{dt} \right) = N \exp \left(-\frac{\omega^2 x^2}{2\kappa T} - \frac{(dx/dt)^2}{2\kappa T} \right) \qquad (8.5)$$

which leads to the following limit values as $t \to \infty$,

$$\left\langle [x\,(t) - \langle x\,(t) \rangle]^2 \right\rangle = \frac{\kappa T}{\omega^2}; \quad \left\langle [v\,(t) - \langle v\,(t) \rangle]^2 \right\rangle = \kappa T. \qquad (8.6)$$

For external noise, the form of $P_{st}\,(x)$, which corresponds to equation (8.1), depends on the properties of the noise. For Ornstein-Uhlenbeck noise (2.6), it is convenient [130] to express $P_{st}\,(x)$ in terms of the following variable

$$q = \varepsilon\,(t) - 2\gamma \frac{dx}{dt} - \frac{2\gamma\tau\omega^2}{1 + 2\gamma\tau} x \qquad (8.7)$$

where $\varepsilon\,(t)$ is the white noise, $\langle \varepsilon\,(t_1)\varepsilon\,(t_2) \rangle = 2\delta\,(t_1 - t_2)$. Then, one gets,

$$P_{st}\left(x, \frac{dx}{dt}, q\right) = N \exp\left\{-\frac{1}{D}\left[\gamma\omega^2\left(1 + \frac{\omega^2\tau^2}{1 + 2\gamma\tau}\right)x^2\right.\right.$$

$$\left.\left. + \gamma\left(1 + 2\gamma\tau + \omega^2\tau^2\right)\left(\frac{dx}{dt}\right)^2 + \frac{\tau\left(1 + 2\gamma\tau\right)}{2}q^2\right]\right\}.$$

$$(8.8)$$

The stationary second moment can easily be obtained from (8.8). For an exponential correlation function (2.6), the variance turns out [131] to be

$$\left\langle [x(t) - \langle x(t)\rangle]^2\right\rangle$$

$$= \frac{2D\tau}{1 - 2\gamma\tau + \omega^2\tau^2}\left\{\frac{\tau^2}{1 + 2\gamma\tau + \omega^2\tau^2}\right.$$

$$\times\left[1 - \left(\frac{(1 + 2\gamma\tau)\sin\omega_1 t}{2\omega_1\tau} + \cos\omega_1 t\right)\exp\left(-\frac{1 + 2\gamma\tau}{2\tau}t\right)\right]$$

$$+ \frac{1 - 2\gamma\tau}{4\omega^2\gamma\tau}[1 - \exp(-2\gamma t)] - \frac{\exp(-2\gamma t)}{4\omega^2}$$

$$\left.\times\left[\frac{1 - 2\gamma\tau}{\omega_1\tau}\sin 2\omega_1 t + \frac{2\gamma - 4\gamma^2\tau + 2\omega^2\tau}{\omega_1^2\tau}\sin^2\omega_1 t\right]\right\}. \quad (8.9)$$

For white noise, this equation reduces to

$$\left\langle [x(t) - \langle x(t)\rangle]^2\right\rangle$$

$$= \frac{D}{2\gamma\omega_1^2}\left\{1 - \left[1 + \frac{\gamma}{\omega_1}\left(\sin 2\omega_1 t + \frac{2\gamma}{\omega_1}\sin^2\omega_1 t\right)\right]\exp(-2\gamma t)\right\}.$$

$$(8.10)$$

The stationary $(t \to \infty)$ values of the variance are the following: for colored noise,

$$\lim_{t\to\infty}\left\langle [x(t) - \langle x(t)\rangle]^2\right\rangle = \frac{D\left(1 + 2\gamma\tau\right)}{2\omega^2\gamma\left(1 + 2\gamma\tau + \omega^2\tau^2\right)} \quad (8.11)$$

and for white noise,

$$\lim_{t \to \infty} \left\langle [x(t) - \langle x(t) \rangle]^2 \right\rangle = \frac{D}{2\omega^2 \gamma}. \tag{8.12}$$

For the following discussion, let us note that the stationary value (8.11) of the variance turns out to be a non-monotonic function of the correlation rate $\lambda = \tau - 1$.

Both equations (8.10) and (8.11) were already obtained in 1945 [132].

It turns out [131], that the mean-squared coordinate and velocity show ballistic behavior at small times, $t \to 0$: for colored noise, both internal and external,

$$\left\langle (\Delta x)^2 \right\rangle \sim t^4 \quad \text{and} \quad \left\langle (\Delta v)^2 \right\rangle \sim t^2 \tag{8.13}$$

while for white noise,

$$\left\langle (\Delta x)^2 \right\rangle \sim t^3 \quad \text{and} \quad \left\langle (\Delta v)^2 \right\rangle \sim t. \tag{8.14}$$

One can also study [131] the change in the time-dependence of the variance for different time intervals depending on the relation between the observation time t and the three characteristic times of the problem: the relaxation time $1/\gamma$, the period of the force-free oscillator $2\pi/\omega$, and the correlation time τ.

8.3   Additive noise and parametric oscillations

In previous chapters we considered linear differential equations with constant coefficients. However, in many practical situations these coefficients are periodical functions of time which leads, in particular, to a Mathieu-type equation

$$\frac{d^2 x}{dt^2} + [a - 2b\cos(2t)] x = \xi(t). \tag{8.15}$$

The properties of the dynamic Mathieu equation ($\xi(t) = 0$ in (8.15)) for different values of a and b are well-known [133][1]. The random force in (8.15) does not influence the first moment $\langle x \rangle$, while for the variance $\sigma^2 \equiv \langle x^2 \rangle - \langle x \rangle^2$ one obtains the following equation [135]

$$\frac{d^3\sigma^2}{dt^3} + 4\left[a - 2b\cos(2t)\right]\frac{d\sigma^2}{dt} + 8b\sin(2t)\,\sigma = 2D \tag{8.16}$$

The solutions of equation (8.16) have basically the same qualitative properties as those of equation (8.15). However, for sufficiently large values of b the variance does not exhibit the expected linear dependence and, instead, increases exponentially with time.

Equation (8.15) has been generalized by including the damping term and the random initial phase in a sinusoidal term [136]. It turns out that for certain parameter values the variance for a damped oscillator is suppressed as compared to its equilibrium value (parametric squeezing).

---

[1] Among the many applications of the Mathieu equation is the dynamic behavior of an acrobat who holds an assistant poised on a pole above his head while he himself stands on a spherical ball rolling on the ground [134]. The analysis of this system, however, requires further experimental investigation, which we leave to the interested reader.

Chapter 9

Nonlinear oscillator with additive noise

9.1 Statistical linearization

The first attempt to solve a nonlinear problem is to replace it (as accurately as possible) by a linear problem which allows an exact solution. For the stochastic differential equation,

$$\frac{d^2x}{dt^2} + \gamma \frac{dx}{dt} + \omega_0^2 x + bf(x) = \sqrt{2D}\xi(t), \tag{9.1}$$

where $bf(x)$ is a nonlinear force, we replace equation (9.1) by the "equivalent" linear equation

$$\frac{d^2x}{dt^2} + \gamma \frac{dx}{dt} + \omega^2 x = \sqrt{2D}\xi(t) \tag{9.2}$$

in such a way that the error

$$\Delta(x) = \omega_0^2 x + bf(x) - \omega^2 x \tag{9.3}$$

is minimal. In the method of statistical linearization [137] ω is chosen to minimize the mean square error $\langle \Delta^2(x) \rangle$ at equilibrium. Thus, ω is the solution of the equation

$$\lim_{t\to\infty} \frac{\partial}{\partial \omega^2} \langle \Delta^2(x) \rangle \equiv \frac{\partial}{\partial \omega^2} \langle \Delta^2(x) \rangle_{as} = 0 \tag{9.4}$$

where the average is performed over all realizations of the random force $\xi(t)$ at equilibrium,

$$\langle \Delta^2(x) \rangle_{as} = \int_{-\infty}^{\infty} \Delta^2(x) P_{as}(x)\, dx. \tag{9.5}$$

It was shown [138] that the procedure (9.4) gives the best approximation for the first two moments, namely, the first asymptotic moment $\langle x(t) \rangle_{as}$ is the same for (9.1) and (9.2), while the difference between the second moments $\langle x^2(t) \rangle_{as}$ is minimized.

On substituting (9.3) into (9.4), one obtains

$$\omega^2 = \omega_0^2 + b \frac{\langle xf(x) \rangle_{as}}{\langle x^2 \rangle_{as}}. \tag{9.6}$$

Then, according to (9.5)-(9.6), the problem is reduced to finding the asymptotic probability distribution $P_{as}(x)$ for the linear equation (9.2). For Gaussian white noise and zero initial conditions,

$$P_{st}(x) = \frac{1}{\sqrt{2\pi\sigma^2}} \exp\left(-\frac{x^2}{2\sigma^2}\right) \tag{9.7}$$

with $\sigma^2 = \langle x^2 \rangle = D/\gamma\omega^2$. If the nonlinear function $f(x)$ is odd in x,

$$f(x) = \sum_{n=1}^{\infty} f_n x^{2n+1} \tag{9.8}$$

then [138]

$$\frac{D}{\gamma\sigma^2} = \omega_0^2 + b \sum_{n=1}^{\infty} f_n c_n \sigma^{2n} \tag{9.9}$$

where the coefficients c_n are given by

$$c_n = \int_{-\infty}^{\infty} x^{2n+2} \exp\left(-\frac{x^2}{2}\right) dx. \tag{9.10}$$

On substituting (9.10) into (9.9), one obtains the equation for the variance σ^2 which generally, has to be solved numerically. The corrections to the outlined process of statistical linearization have been considered in [138].

9.2   Double-well oscillator with additive noise

The motion of a particle in a double-well potential $U = -\frac{\omega^2 x^2}{2} + \frac{bx^4}{4}$ is described by the following equation

$$\frac{d^2x}{dt^2} + 2\gamma\frac{dx}{dt} = -\frac{dU}{dx} + \sqrt{2D}\xi(t) = \omega^2 x - bx^3 + \sqrt{2D}\xi(t) \quad (9.11)$$

which is a nonlinear generalization of (8.1).

A detailed analysis has been carried out [139] of the correlation function $\langle x(t) x(0) \rangle$ and its Fourier component $Q(\Omega)$ (power spectrum) defined as

$$Q(\Omega) = \frac{1}{\pi}\text{Re}\int_0^\infty dt \, \exp(\Omega t)\langle x(t) x(0)\rangle, \quad (9.12)$$

In terms of the dimensionless "time" ωt and "coordinate" $\frac{\sqrt{b}}{\omega}x$, two characteristic dimensionless parameters are γ/ω and β, where

$$\beta = \frac{bD}{2\omega^4} = \frac{D}{8\Delta U} \quad (9.13)$$

with ΔU being the height of the potential barrier, $\Delta U = \omega^4/4b$.

It turns out [139] that for the underdamped oscillator, $\gamma/\omega \ll 1$, the function $Q(\omega)$ has either two or three distinct peaks, depending on the noise intensity β. One of the peaks is located at frequency $\Omega = 0$. For $\beta \ll 1$, this peak is related to the fluctuational transitions between the two stable points of the potential. The intensity of this peak is of order of ω^2/b, and its half-width is exponentially small, being proportional to $\gamma\exp\left[-(4\beta)^{-1}\right]$. Another peak at $\beta \ll 1$ is related to the small amplitude intrawell vibrations, and this peak is located at $\Omega \simeq \sqrt{2}\omega$. With increasing noise intensity, an increasingly important role is played by the third peak related to the overbarrier vibrations and located at $\Omega \simeq \pi\omega/\ln\left(16\beta^{-1}\right)$ for $\ln\left(16\beta^{-1}\right) \gg 1$. The detailed calculations of shapes, intensities and locations of the peaks have been supported by analogue experiments using an electronic circuit [139].

9.3   Double-well oscillator driven by two periodic fields (vibrational resonance)

As we have already seen in section 5.1, a shift of the stable points occurs either through multiplicative noise or through a parametric periodic force even though the physical mechanism is quite different. Indeed, this effect is determined by the low frequencies of the correlator of noise, and by the high frequencies of an external field. It turns out [140] that the analogous effect occurs for two additive periodic fields as well. A bistable underdamped oscillator subject to two periodic field is described by the following equation

$$\frac{d^2x}{dt^2} + \gamma\frac{dx}{dt} - \omega_0^2 x + bx^3 = A\sin\left(\omega t\right) + C\sin\left(\Omega t\right). \qquad (9.14)$$

Suppose now that one of the fields has large amplitude (larger than the barrier height $\Delta U = \omega_0^2/4b$), $C > \omega_0^2/4b$, and high frequency, $\Omega \gg \omega$. The former means that during each half-period, this field transfers the system from one potential well to the other. A similar situation holds in a random system where the large amplitude field in (9.14) is replaced by an additive random force, which plays the same role of switching a system between the two minima. Therefore, by choosing the appropriate relation between the input signal $A\sin\left(\omega t\right)$ and the amplitude C of the large signal (or the strength of noise), one can obtain a non-monotonic dependence of the output signal on the noise strength (stochastic resonance considered in section 7.5) or on the amplitude C (vibrational resonance [140]).

For the qualitative description of vibrational resonance, consider equation (9.14) [84]. Due to the high frequency Ω, one has two different time scales. Therefore, let us seek a solution of equation (9.14) in the form

$$x\left(t\right) = y\left(t\right) - \frac{C\sin\left(\Omega t\right)}{\Omega^2}. \qquad (9.15)$$

The first term on the right-hand side will be assumed to vary significantly only over times of t, while the second one varies rapidly. On substituting (9.15) into (9.14), one can average over a single cycle time of $\sin\left(\Omega t\right)$. All odd powers of $\sin\left(\Omega t\right)$ vanish upon averaging, while the $\sin^2(\Omega t)$ term gives $1/2$. Finally, one obtains the following equation for $X(t)$, the mean value of $y(t)$, $X(t) = \langle y(t) \rangle$,

$$\frac{d^2 X}{dt^2} + \gamma \frac{dX}{dt} - \left(\omega_0^2 - \frac{3bC^2}{2\Omega^4} \right) X + bX^3 = A \sin(\omega t) \qquad (9.16)$$

where the slowly-varying term $A \sin(\omega t)$ does not change under averaging over short times.

In order to solve equation (9.16), one needs some additional assumptions which could be weak, fast or slow driving [141]. We perform the simplest procedure of linearization of a nonlinear oscillator [46] similar to stochastic linearization considered in the previous section, seeking the solution of (9.16) of the form

$$X(t) \approx \Theta \sin(\omega t - \theta). \qquad (9.17)$$

Retaining only the first term in a Fourier series of the nonlinear term in (9.16) and averaging over the period $2\pi/\omega$ of the external field, the bX^3 term can be replaced by $3b\Theta^2 X/4$. Equation (9.16) then reduces to

$$\frac{d^2 X}{dt^2} + \gamma \frac{dX}{dt} + \varpi^2 X = A \sin(\omega t) \qquad (9.18)$$

with the renormalized frequency

$$\varpi = \sqrt{\frac{3b\Theta^2}{4} + \frac{3bC^2}{2\Omega^4} - \omega_0^2}. \qquad (9.19)$$

Equation (9.16) is identically satisfied by (9.17) if

$$\gamma \omega \Theta = A \sin(\theta) \text{ and } \frac{3b\Theta^3}{4} - \left(\omega_0^2 - \frac{3bC^2}{2\Omega^4} + \omega^2 \right) \Theta = A \cos(\theta). \qquad (9.20)$$

Eliminating Θ from equations (9.19) and (9.20), one obtains

$$\varpi^2 = \frac{3bC^2}{2\Omega^4} - \omega_0^2 + \frac{3bA^2}{4 \left[\gamma^2 \omega^2 + (\varpi^2 - \omega^2)^2 \right]}. \qquad (9.21)$$

A resonance in the linear equation (9.18) occurs when $\varpi = \omega$. Substituting the latter into equation (9.21), one can find the relation between the amplitudes and frequencies of the two driving fields in equation (9.14) which produce the resonance behavior. This condition has the form

$$\omega^2 = \frac{3bC^2}{2\Omega^4} + \frac{3bA^2}{4\gamma^2\omega^2} - \omega_0^2. \tag{9.22}$$

In addition to the approximate solution of equation (9.16), one can perform a quantitative analysis of this equation [142]. Equation (9.16) describes the driven motion in the effective potential U_{eff} of the form

$$U_{eff} = \left(\omega_0^2 - \frac{3bC^2}{2\Omega^4}\right)\frac{X^2}{2} - \frac{bX^4}{4}. \tag{9.23}$$

Hence, the locations of new equilibrium states about which slow oscillations are executed, depend on the parameter $3bC^2/2\omega_0^2\Omega^4$. If this parameter is smaller than unity, $\left(3bC^2/2\omega_0^2\Omega^4\right) < 1$, there are two equilibrium states

$$X_{1,2} = \pm\sqrt{\frac{\omega_0^2}{b} - \frac{3C^2}{\Omega^4}}. \tag{9.24}$$

For $\left(3bC^2/2\omega_0^2\Omega^4\right) > 1$, there is only one equilibrium state, $X = 0$. Thus, transition from bimodal to unimodal distribution depends on the amplitude of the high-frequency field (analogous to the shift of the stable points induced by noise, considered in section 5.1).

The equation for the deviation, $Y = X - X_{1,2}$, of X from one of the stable equilibrium states $X_{1,2}$, is obtained by the substitution $X = Y + X_{1,2}$ into equation (9.16),

$$\frac{d^2Y}{dt^2} + \gamma\frac{dY}{dt} + 2\left(\omega_0^2 - \frac{3bC^2}{2\Omega^4}\right)Y + 3bX_{1,2}Y^2 + bY^3 = A\sin(\omega t), \tag{9.25}$$

while the derivation from $X = 0$ is described by (9.16). We see that the resonance frequency of the system

$$\omega_{res,1} = \sqrt{2\left(\omega_0^2 - \frac{3bC^2}{2\Omega^4}\right)} \tag{9.26}$$

decreases from $\sqrt{2}\omega_0$ to zero as the amplitude C increases from zero to $C_0 = \sqrt{2\omega_0^2\Omega^4/3b}$.

On the other hand, for $X = 0$, the resonance frequency

$$\omega_{res,2} = \sqrt{\frac{3bC^2}{2\Omega^4} - \omega_0^2} \qquad (9.27)$$

increases with an increase in C from C_0. Therefore, one concludes that the resonance frequency is a non-monotonic function of C. Numerical solutions of linearized equations (9.25) for $C < C_0$ and (9.16) for $C > C_0$, show [142] that the amplitude of the output signal has a maximum as a function of the amplitude C of the high frequency field. Moreover, the *bona fide* resonance occurs as a function of the high - frequency Ω when the low frequency ω is fixed, as well as being a function of ω at fixed Ω.

All the above results have been obtained for an underdamped oscillator. It turns out [142] that similar effects also take place for an overdamped oscillator. Experimental evidence for vibrational resonance has been obtained in an optical system [143] and in an electronic circuit [144].

Chapter 10

Harmonic oscillator with random frequency

10.1 First moment for the random frequency

10.1.1 *Force-free oscillator*

The equation of motion of the oscillator with a random frequency

$$\frac{d^2x}{dt^2} + 2\gamma\frac{dx}{dt} + \omega^2\left[1 + \xi\left(t\right)\right]x = 0 \tag{10.1}$$

has been studied extensively (The comprehensive list of references can be found in [145]). We list here the main results for white and colored noise.

10.1.2 *White noise*

It turns out that the fluctuations of frequency do not effect the first moment of the underdamped oscillator provided the fluctuations are delta correlated, and $\langle x\left(t\right)\rangle$ remains equal to noise-free solution $x\left(t\right)$ defined in (1.4).

10.1.3 *Colored noise*

To get the equation for the averaged first moment from equation (10.1) with colored noise, one has to use some approximate methods. The approximations are different for slow and fast fluctuations where the correlation rate is much smaller or larger than the characteristic time of the deterministic system. For the slow (adiabatic) fluctuations, one uses an expansion in small parameter $\sigma^2/\omega_0 < 1$ which leads to [146]

$$\frac{d^2 \langle x \rangle}{dt^2} + 2\gamma \left(1 - \langle \xi^2 \rangle\right) \frac{d \langle x \rangle}{dt}$$

$$+ \left(\omega^2 + \omega^2 \langle \xi^2 \rangle + 2\gamma \left\langle \xi \frac{d\xi}{dt} \right\rangle\right) \langle x \rangle = 0. \qquad (10.2)$$

For the fast fluctuations the small parameter is $\sigma^2 \tau < 1$, and the equation is of the form [28]

$$\frac{d^2 \langle x \rangle}{dt^2} + \left(2\gamma + \omega^2 c_2\right) \frac{d \langle x \rangle}{dt} + \omega^2 \left(1 - \omega c_1\right) \langle x \rangle = 0 \qquad (10.3)$$

where the coefficients

$$c_1 = 2 \int_0^\infty \langle \xi(t) \xi(t - \tau) \rangle \sin(2\omega\tau) \, d\tau,$$

$$c_2 = 2 \int_0^\infty \langle \xi(t) \xi(t - \tau) \rangle \left[1 - \cos(2\omega\tau)\right] d\tau. \qquad (10.4)$$

In agreement with the foregoing these coefficients vanish for the white noise correlators. Equations (10.2) and (10.3) show that the fluctuations in frequency cause a gain in $\langle x \rangle$ for adiabatic fluctuations and damping for fast fluctuations.

Stability conditions of equations (10.1) have been considered [147] also for a signal-modulated colored noise with the correlator (2.7). It turned out [147] that the maximum stability of power spectrum $S(\Omega)$ occurs at "resonance" frequencies $\Omega = 2\omega_2/n$. Numerical analysis shows that for large correlation rate λ, the main resonance ($n = 1$) appears at $\Omega = 2\omega_2$, while for small λ, one sees additional resonances ($n = 2, 3$) at $\Omega = \omega_2$ and $\Omega = 2\omega_2/3$. Similar maxima arise for equation (10.1) with random force $\xi(t)$ replaced by the periodic force $A \cos(\omega_2 t)$.

A comparison between different forms of noise (white, Ornstein-Uhlenbeck and signal-modulated) in equation (10.1) has been performed in [148]. Moreover, in [148] and [149] the simultaneous action of multiplicative periodic signal and noise has been considered in detail. In the absence of noise, the well-known phenomenon of parametric instabilities occurs. The effect of noise on parametric instability is twofold: it triggers the instability before its deterministic onset, and for a nonlinear system it inhibits the nonlinearly saturated oscillatory response above the deterministic onset. These effects have been observed experimentally for Faraday waves (motion of the free surface of a fluid parametrically excited by an oscillatory and random external forces)[148], [149], and electronic circuit [149].

10.2 Driven oscillator

The second-order differential equation for driven harmonic oscillation with random frequency can be written as two first-order differential equations

$$\frac{dx}{dt} = y, \tag{10.5}$$

$$\frac{dy}{dt} = -2\gamma y - \omega^2 x - \omega^2 \xi x + A \sin \Omega t \tag{10.6}$$

which, after averaging, take the following form:

$$\frac{d}{dt} \langle x \rangle = \langle y \rangle \tag{10.7}$$

$$\frac{d}{dt} \langle y \rangle = -2\gamma \langle y \rangle - \omega^2 \langle x \rangle - \omega^2 \langle \xi(t) x \rangle + A \sin (\Omega t). \tag{10.8}$$

The new correlator $\langle \xi(t) x \rangle$ has to be found separately. To this end, we use the Shapiro-Loginov procedure (2.8). Multiplying Eq. (10.5) by ξ, one gets after averaging,

$$\left\langle \xi(t) \frac{dx}{dt} \right\rangle = \langle \xi(t) y \rangle. \tag{10.9}$$

Inserting (10.9) into (2.8) results in

$$\frac{d \langle \xi(t) x \rangle}{dt} = \langle \xi(t) y \rangle - \lambda \langle \xi(t) x \rangle. \tag{10.10}$$

Using the procedure analogous to (10.9)-(10.10) for the correlator $\langle \xi(t) y \rangle$, one gets

$$\frac{d \langle \xi(t) y \rangle}{dt} = \left\langle \xi(t) \frac{dy}{dt} \right\rangle - \lambda \langle \xi(t) y \rangle. \tag{10.11}$$

Multiplying Eq. (10.6) by ξ and averaging, one obtains

$$\left\langle \xi \frac{dy}{dt} \right\rangle = -2\gamma \langle \xi y \rangle - \omega^2 \langle \xi x \rangle - \omega^2 \langle \xi^2 x \rangle. \tag{10.12}$$

Equation (10.12) contains the higher order correlator $\langle \xi^2 x \rangle$, and one has to use a decoupling procedure. For dichotomous noise $\xi^2 = \sigma^2$, and Eq. (10.12) can be rewritten as

$$\left\langle \xi \frac{dy}{dt} \right\rangle = -2\gamma \langle \xi y \rangle - \omega^2 \langle \xi x \rangle - \omega^2 \sigma^2 \langle x \rangle. \tag{10.13}$$

Inserting (10.13) into (10.11) results in

$$\frac{d \langle \xi(t) y \rangle}{dt} = -2\gamma \langle \xi y \rangle - \omega^2 \langle \xi(t) x \rangle - \omega^2 \sigma^2 \langle x \rangle - \lambda \langle \xi(t) y \rangle. \tag{10.14}$$

We thus obtain a system of four equations, (10.7), (10.8), (10.10) and (10.14), for four variables, $\langle x \rangle$, $\langle y \rangle$, $\langle \xi x \rangle$, and $\langle \xi y \rangle$.

From these equations one can easily find the fourth-order differential equation for $\langle x \rangle$

$$\frac{d^4 \langle x \rangle}{dt^4} + 2(\lambda + 2\gamma) \frac{d^3 \langle x \rangle}{dt^3} + \left(2\omega^2 + \lambda^2 + 6\lambda\gamma + 4\gamma^2\right) \frac{d^2 \langle x \rangle}{dt^2}$$

$$+ \left[2\omega^2 (\lambda + 2\gamma) + 2\lambda\gamma (\lambda + 2\gamma)\right] \frac{d \langle x \rangle}{dt}$$

$$+ \left[\omega^2 (\omega^2 + \lambda^2 + 2\lambda\gamma) - \omega^2 \sigma^2\right] \langle x \rangle$$

$$= \left(\omega^2 + \lambda^2 + 2\lambda\gamma - \Omega^2\right) A \sin(\Omega t) + 2(\lambda + \gamma) A\Omega \cos(\Omega t). \tag{10.15}$$

We seek the solution of Eq. (10.15) in the form

$$\langle x \rangle = \langle x \rangle_0 + \langle x \rangle_A \tag{10.16}$$

where the output signal $\langle x \rangle_A$ is induced by an external field $A \sin(\Omega t)$ and $\langle x \rangle_0$ is defined by internal dynamics. For purposes of this discussion we ignore the possible instability of an underdamped oscillator for adiabatic fluctuations [146].

Let us write the solution $\langle x \rangle_A$ of Eq. (10.15) in the form

$$\langle x \rangle_A = a \sin(\Omega t + \phi) \tag{10.17}$$

Then, one easily finds [14]

$$a = \left[\frac{f_1^2 + f_2^2}{f_3^2 + f_4^2} \right]^{\frac{1}{2}}, \quad \phi = \tan^{-1} \left(\frac{f_1 f_3 - f_2 f_4}{f_1 f_4 + f_2 f_3} \right) \tag{10.18}$$

where

$$f_1 = 2(\lambda + \gamma) A\Omega, \quad f_2 = \left(\Omega^2 - \omega^2 - \lambda^2 - 2\lambda\gamma \right) A$$

$$f_3 = 2\left(\Omega^2 - \omega^2 \right)\left(\Omega^2 - \omega^2 - \lambda^2 \right) - \omega^2\sigma^2 - \left(6\lambda\gamma + 4\gamma^2 \right)\Omega^2 + 2\lambda\gamma\omega^2,$$

$$f_4 = \Omega(\lambda + 2\gamma)\left[\left(\omega^2 - \Omega^2 \right) + \lambda\gamma \right]. \tag{10.19}$$

In the absence of friction, $\gamma = 0$, equations (10.18)-(10.19) take the following form

$$a = A\sqrt{\frac{\left[\left(\Omega^2 - \omega^2 - \lambda^2 \right)^2 + 4\lambda^2\Omega^2 \right]}{\left[\left(\Omega^2 - \omega^2 \right)\left(\Omega^2 - \omega^2 - \lambda^2 \right) - \omega^2\sigma^2 \right]^2 + 4\lambda^2\Omega^2 \left(\Omega^2 - \omega^2 \right)^2}} \tag{10.20}$$

and

$$\phi = \tan^{-1} \frac{2\Omega\lambda\omega^2\sigma^2}{\left(\Omega^2 - \omega^2 \right)\left[\left(\Omega^2 - \omega^2 - \lambda^2 \right)^2 + 4\Omega^2\lambda^2 \right] - \left(\Omega^2 - \omega^2 - \lambda^2 \right)\omega^2\sigma^2}. \tag{10.21}$$

For small σ^2 Eq. (10.20) reduces to Eq. (8.6) of [150] found in a different context by perturbation theory.

Before analyzing Eqs. (10.20)-(10.21), let us consider the limiting case of white Gaussian noise which corresponds to $\sigma^2 \to \infty$ and

$\lambda \to \infty$ with a constant ratio. Then,

$$a = \frac{A}{(\omega^2 - \Omega^2)}; \quad \phi = 0 \qquad (10.22)$$

as expected, since for white noise, $\langle x \rangle$ satisfies the following equation [28]

$$\frac{d^2 \langle x \rangle}{dt^2} + \omega^2 \langle x \rangle = A \sin \Omega t. \qquad (10.23)$$

However, for colored noise (dichotomous in the present case) the output signal (10.20) shows non-monotonic dependence on the noise strength σ^2 and on the correlation rate λ (stochastic resonance). Indeed, the amplitude of the output signal A reaches a maximum at

$$\sigma^2 = \left(\Omega^2 - \omega^2\right)\left(\Omega^2 - \omega^2 - \lambda^2\right)/\omega^2. \qquad (10.24)$$

In Fig. 10.1 we show the dependence of the amplitude of a stationary signal a on the correlation rate λ for $A = \sigma^2 = \omega = 1, \gamma = 0$ and different frequencies Ω of the external field. This graph shows typical SR non-monotonic behavior for $\Omega < \omega$. With a rise in Ω the positions of the maxima are shifted to higher λ, while the heights are non-monotonic functions of Ω.

Note that the resonance amplitude of an underdamped harmonic oscillator (Eq. (10.20) with $\Omega = \omega$) remains restricted in the presence of colored noise ("effective damping").

10.3   Second moment for a random frequency

In the absence of an external field there are different ways to calculate the second moment of equation (10.1). We bring here without proof only a few results. Firstly, the second moment of the oscillator with random frequency subjected to dichotomous multiplicative noise $\xi(t)$ of strength σ^2 and correlation rate λ, and additive white noise $\eta(t)$ defined as $\langle \eta(t_1)\eta(t_2)\rangle = 2\alpha\delta(t_1 - t_2)$ is given by [151]

$$\langle x^2 \rangle = \alpha \left[4\gamma\omega^2 - \frac{2\omega^4\sigma^2(\lambda + 4\gamma)^2}{(\lambda + 2\gamma)[4\omega^2 + \lambda(\lambda + 4\gamma)]} \right]^{-1} \qquad (10.25)$$

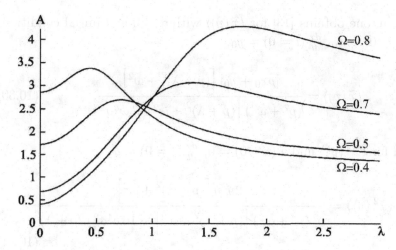

Fig. 10.1 The amplitude of a stationary signal as a function of the correlation rate for parameters $a = \sigma^2 = \omega = 1$ and $\gamma = 0$. The curves displayed correspond to different values of the frequency of an external field. [Reprinted figure with permission from Ref. 14. Copyright (2003) by the American Physical Society.]

which for the white noise reduces to

$$\langle x^2 \rangle = \frac{\alpha}{2\omega^2 \left(2\gamma - D\omega^2\right)} \qquad (10.26)$$

showing the well known "energetic" instability for $D\omega^2 > 2\gamma$.

Analogous calculations for equation

$$\frac{d^2x}{dt^2} + \gamma \frac{dx}{dt} + \omega^2 \left[1 + \xi(t)\right] x = f(x) \qquad (10.27)$$

with white noise $f(x)$, $\langle f(x_1) f(x_2) \rangle = 2\alpha\delta(x_1 - x_2)$, and Poisson noise $\xi(t)$ defined by Poisson parameter λ and jump length S, gives for a second moment [152]

$$\langle x^2 \rangle = \frac{\alpha\omega^2}{\gamma\omega^2 - \frac{\lambda S^2}{2}}, \qquad (10.28)$$

i.e., the instability occurs at $\lambda S^2 > 2\gamma\omega^2$.

Another way to find the first and second moments is the use of the Laplace transformation

$$\langle \tilde{x}(p) \rangle = \int_0^\infty \langle x(t) \rangle \exp\left(-pt\right) dt \qquad (10.29)$$

Then, one obtains [13] for $\langle \widetilde{x}\,(p) \rangle$ with $\gamma = 0$ and initial conditions $x\,(t = 0) = x_0$, $\frac{dx}{dt}\,(t = 0) = y_0$,

$$\langle \widetilde{x}\,(p) \rangle = \frac{[px_0 + y_0]\left[(p+\lambda)^2 + \omega^2\right]}{(p^2 + \omega^2)\left[(p+\lambda)^2 + \omega^2\right] - \omega^4 \sigma^2}, \tag{10.30}$$

and for $\langle \widetilde{x}^2\,(p) \rangle$ and $x\,(t = 0) = 0$, $\frac{dx}{dt}\,(t = 0) = y_0$

$$\langle \widetilde{x}^2\,(p) \rangle = \frac{2y_0^2 p \left[(p+\lambda)^2 + \omega^2\right]}{p^2 (p^2 + \omega^2)\left[(p+\lambda)^2 + \omega^2\right] - 16\omega^4 \sigma^2 (2p+\lambda)^2} \tag{10.31}$$

10.4   Maxwell equation with random dielectric constant

We replace equation (10.1) with $\gamma = 0$ by the isomorphic Maxwell equation with an external force, substituting the electric field E and coordinate x instead of the coordinate x and time t,

$$\frac{d^2 E}{dx^2} + [\epsilon_0 + \epsilon_1 \xi\,(x)]\,E = A\sin(kx) \tag{10.32}$$

where for white noise

$$\langle \xi\,(x)\,\xi\,(x_1) \rangle = 2D\delta\,(x - x_1) \tag{10.33}$$

and for dichotomous noise

$$\langle \xi\,(x)\,\xi\,(x_1) \rangle = \sigma^2 \exp\,(-\lambda\,|x - x_1|). \tag{10.34}$$

There are also other models, isomorphic to that described by equation (10.1), such as the theory of paramagnetic resonance, high-dimensional Hamiltonian systems [153] , and Anderson localization in solids for which equation (10.1) is the one-dimensional Shrödinger equation for a single particle in a delta-correlated potential $\xi\,(t)$ [154].

The second-order differential equation (10.32) can be rewritten as two first-order differential equations,

$$\frac{dE}{dx} = y,$$

$$\frac{dy}{dx} = -\epsilon_0 E - \epsilon_1 \xi(x) E + A \sin(kx).$$

(10.35)

Equation (10.35) is similar to (10.5)-(10.6). The calculations quite similar to (10.7)-(10.15), lead to the following fourth-order differential equation for $\langle E \rangle$

$$\frac{d^4 \langle E \rangle}{dx^4} + 2\lambda \frac{d^3 \langle E \rangle}{dx^3} + \left(2\epsilon_0 + \lambda^2\right) \frac{d^2 \langle E \rangle}{dx^2} + 2\epsilon_0 \lambda \frac{d \langle E \rangle}{dx}$$

(10.36)

$$+ \left[\epsilon_0 \left(\epsilon_0 + \lambda^2\right) - \epsilon_1 \sigma^2\right] \langle E \rangle = \left(\epsilon_0 + \lambda^2 - k^2\right) A \sin kx + 2\lambda A k \cos kx.$$

The response to an external field is defined by the solution of the non-homogeneous equation (10.36) which has the following form

$$\langle E \rangle_A = a \sin(kx + \varphi)$$

(10.37)

with

$$a = A \left[\frac{\left[\left(k^2 - \epsilon_0 - \lambda^2\right)^2 + 4\lambda^2 k^2\right]}{\left[\left(k^2 - \epsilon_0\right)\left(k^2 - \epsilon_0 - \lambda^2\right) - \epsilon_1^2 \sigma^2\right]^2 + 4\lambda^2 k^2 \left(k^2 - \epsilon_0\right)^2} \right]^{\frac{1}{2}},$$

$$\tan \varphi = \frac{2k\lambda \epsilon_1 \sigma^2}{\left(k^2 - \epsilon_0\right)\left[\left(k^2 - \epsilon_0 - \lambda^2\right)^2 + 4k^2 \lambda^2\right] - \left(k^2 - \epsilon_0 - \lambda^2\right) \epsilon_1^2 \sigma^2}.$$

(10.38)

Analogously, one obtains

$$\langle y \rangle_A = ak \cos(kx + \phi), \quad \langle \xi E \rangle_A = L \sin(kx + \phi) + M \cos(kx + \phi),$$

$$\langle \xi y \rangle_A = (\lambda L - kM) \sin(kx + \phi) + (\lambda M + kL) \cos(kx + \phi)$$

(10.39)

where

$$L = \frac{k^2 - \epsilon_0 - \lambda^2}{2\lambda k} M, \qquad M = \frac{2ak\lambda\epsilon_1^2\sigma^2}{\left[(k^2 - \epsilon_0 - \lambda^2)^2 + 4k^2\lambda^2\right]}. \quad (10.40)$$

To find the second moments, one multiplies equations (10.35) by $2E$ and $2y$, respectively, which gives

$$\frac{d}{dx}\langle E^2 \rangle = 2\langle Ey \rangle,$$

$$\frac{d}{dx}\langle y^2 \rangle = -2\epsilon_0 \langle Ey \rangle - 2\epsilon_1 \langle \xi Ey \rangle + 2A\langle y \rangle \sin kx. \quad (10.41)$$

Analogously, multiplying equation (10.35) by y and E, and summarizing these equations, results in

$$\frac{d}{dx}\langle Ey \rangle = \langle y^2 \rangle - \epsilon_0 \langle E^2 \rangle - \epsilon_1 \langle E^2\xi \rangle + A\langle E \rangle \sin kx. \quad (10.42)$$

Equation (10.42) contains a new correlator $\langle E^2\xi \rangle$. One can calculate this and the analogous correlator $\langle y^2\xi \rangle$ using the Shapiro-Loginov approximation (2.8), which leads to

$$\frac{d}{dx}\langle E^2\xi \rangle = 2\langle Ey\xi \rangle - \lambda\langle E^2\xi \rangle,$$

$$\frac{d}{dx}\langle y^2\xi \rangle = -2\epsilon_0 \langle Ey\xi \rangle - 2\epsilon_1^2\sigma^2 \langle Ey \rangle - \lambda\langle y^2\xi \rangle + 2A\langle y\xi \rangle \sin kx.$$

$$(10.43)$$

Repeating again the Shapiro-Loginov procedure for the correlator $\langle Ey\xi \rangle$, one gets finally,

$$\frac{d}{dx}\langle Ey\xi \rangle = \langle y^2\xi \rangle - \epsilon_0 \langle E^2\xi \rangle - \epsilon_1^2\sigma^2 \langle E^2 \rangle - \lambda\langle Ey\xi \rangle + a\langle E\xi \rangle \sin kx.$$

$$(10.44)$$

In obtaining (10.43) and (10.44) we used the assumption of dichotomous noise. Thus, just as in the transition from (11.18) to (11.20), one can replace ξ^2 by σ^2 inside the averages.

Six equations (10.41)-(10.44) can be reduced to the following two third-order differential equations for $\langle E^2 \rangle$ and $\langle E^2 \xi \rangle$,

$$\frac{d^3}{dx^3} \langle E^2 \rangle + 4\epsilon_0 \frac{d}{dx} \langle E^2 \rangle + 4\epsilon_1 \frac{d}{dx} \langle E^2 \xi \rangle + 2\lambda\epsilon_1 \langle E^2 \xi \rangle$$

$$= 4A \langle y \rangle \sin kx + 2A \frac{d}{dx} \left(\langle E \rangle \sin kx \right), \qquad (10.45)$$

$$\frac{d^3}{dx^3} \langle E^2 \xi \rangle + 3\lambda \frac{d^2}{dx^2} \langle E^2 \xi \rangle + \left(4\epsilon_0 + 3\lambda^2\right) \frac{d}{dx} \langle E^2 \xi \rangle$$

$$+ \lambda \left(4\epsilon_0 + \lambda^2\right) \langle E^2 \xi \rangle + 4\epsilon_1^2 \sigma^2 \frac{d}{dx} \langle E^2 \rangle + 2\lambda\epsilon_1^2 \sigma^2 \langle E^2 \rangle$$

$$= 2A\lambda \langle E\xi \rangle \sin kx + 4A \langle y\xi \rangle \sin kx$$

$$+ 2A \frac{d}{dx} \left(\langle E\xi \rangle \sin kx \right), \qquad (10.46)$$

where $\langle E \rangle$, $\langle y \rangle$, $\langle E\xi \rangle$, and $\langle y\xi \rangle$ are given in (10.37)-(10.39).

One can proceed now as in the calculation of the first moments, by excluding the correlator $\langle E^2 \xi \rangle$ from Eqs. (10.45)-(10.46) and by solving the obtained six order differential equation for $\langle E^2 \rangle$, but we do not include here these cumbersome formulas.

10.4.1 *Driven Maxwell equation*

Let us consider now response to an external field. To this end, we put the zero boundary conditions in equations (10.45)-(10.46). Moreover, we will not write the cumbersome formulas for the Laplace transform of the right hand side of these equation, restricting ourselves only to the asymptotic solutions at $x \to \infty$ ($p \to 0$). Then, equations (10.45)-(10.46) are reduced to

$$2\lambda\epsilon_1^2 \sigma^2 \left\langle E^2 \right\rangle_{as} + \lambda \left(\lambda^2 + 4\epsilon_0\right) \left\langle E^2 \xi \right\rangle_{as} = 2A \left[(\lambda \langle E\xi \rangle + 2 \langle y\xi \rangle) \sin kx \right]_{as},$$

$$2\lambda\epsilon_1 \left\langle E^2 \xi \right\rangle_{as} = 4A \left(\langle y \rangle \sin kx \right)_{as}$$

$$\qquad (10.47)$$

which shows [15] that the asymptotic value of the second moment $\left\langle E^2 \right\rangle_{as}$, induced by the external field, is a non-monotonic function of σ^2 and λ. Thus, the energy of the output signal, which is proportional to $\left\langle E^2 \right\rangle_{as}$, depends non-monotonically both on σ^2 and λ.

Chapter 11

Harmonic oscillator with random damping

11.1 First moment for random damping

11.1.1 *Force-free oscillator*

Let us first consider the free motion of an oscillator, rewriting for this purpose equation (1.14) with $m = 1$, and $A = 0$, as

$$\frac{d^2x}{dt^2} + 2\gamma\left[1 + \xi\left(t\right)\right]\frac{dx}{dt} + \omega^2x = 0 \tag{11.1}$$

or, in the equivalent form,

$$L\{x\} = -2\gamma\xi\frac{dx}{dt}; \quad L\{x\} \equiv \left(\frac{d^2}{dt^2} + 2\gamma\frac{d}{dt} + \omega^2\right)x. \tag{11.2}$$

In order to convert the differential equation (11.1) into an integro-differential equation we apply, following [151], the operator L^{-1} to Eq. (11.2) which gives

$$x = -L^{-1}\left\{2\gamma\xi\frac{dx}{dt}\right\}. \tag{11.3}$$

Using the fact that $L\left[L^{-1}\{f\}\right] \equiv f$, one can easily check that the integral operator L^{-1} inverse to the differential operator L defined in (11.2) has the following form

$$L^{-1}\{f\} \equiv \frac{1}{\omega_1} \int_0^t dt_1 \exp\left[-\gamma(t-t_1)\right] \sin\left[\omega_1(t-t_1)\right] f(t_1);$$

$$\omega_1 = \sqrt{\omega^2 - \gamma^2},$$

$$(11.4)$$

i.e., according to (11.2),

$$x(t) = -\frac{2\gamma}{\omega_1} \int_0^t dt_1 \exp\left[-\gamma(t-t_1)\right] \sin\left[\omega_1(t-t_1)\right] \xi(t_1) \frac{dx}{dt}(t_1)$$

$$(11.5)$$

and

$$\frac{dx}{dt} = \frac{2\gamma}{\omega_1} \int_0^t dt_1 \exp\left[-\gamma(t-t_1)\right] \xi(t_1) \frac{dx}{dt}(t_1) \left\{\gamma \sin\left[\omega_1(t-t_1)\right]\right.$$

$$\left. - \omega_1 \cos\left[\omega_1(t-t_1)\right]\right\}.$$

$$(11.6)$$

On substituting (11.6) into the right hand side of equation (11.2), one obtains

$$\left(\frac{d^2}{dt^2} + 2\gamma\frac{d}{dt} + \omega^2\right) x(t)$$

$$= -\frac{4\gamma^2}{\omega_1} \int_0^t dt_1 \exp\left[-\gamma(t-t_1)\right] \xi(t)\xi(t_1)$$

$$\times \frac{dx}{dt}(t_1) \left\{\gamma \sin\left[\omega_1(t-t_1)\right] - \omega_1 \cos\left[\omega_1(t-t_1)\right]\right\}. \quad (11.7)$$

On averaging (11.7) for noise one can use the simplest version of the splitting of averages [151]

$$\left\langle \xi(t)\xi(t_1)\frac{dx}{dt}(t_1)\right\rangle = \langle \xi(t)\xi(t_1)\rangle \left\langle \frac{dx}{dt}(t_1)\right\rangle. \quad (11.8)$$

The substitution of (11.8) into the averaging equation (11.7) shows that for white noise (2.4), one gets

$$\left[\frac{d^2}{dt^2} + 2\gamma(1 - 2\gamma D)\frac{d}{dt} + \omega^2\right] \langle x \rangle = 0, \quad (11.9)$$

i.e., the presence of white noise in the original equation (11.1) leads to a decrease of damping if $2\gamma D < 1$. Moreover, if $2\gamma D > 1$, i.e., if the noise is sufficiently strong, the effective damping becomes negative, so that the average value of the coordinate x increases in time, which indicates an instability. On the other hand, for the exponentially correlated noise (2.5) one gets [24]

$$\frac{d^2 \langle x \rangle}{dt^2} + 2\gamma \frac{d \langle x \rangle}{dt} + \frac{4\gamma^2 \sigma^2}{\omega_1} \int_0^t \exp\left[-(\lambda+\gamma)(t-t_1)\right] \{\gamma \sin\left[\omega_1 (t-t_1)\right]$$

$$- \omega_1 \cos\left[\omega_1 (t-t_1)\right]\} \frac{d \langle x \rangle}{dt} (t_1)\, dt_1 + \omega^2 \langle x \rangle = 0. \tag{11.10}$$

Application of the Laplace transform (10.29) to equation (11.10) yields

$$\frac{(p^2+\omega^2+2\gamma p)\left[(p+\lambda)(p+\lambda+2\gamma)+\omega^2\right]-4\sigma^2 p\gamma^2 (p+\lambda)}{(p+\lambda+2\gamma)(p+\lambda)+\omega^2} \langle \tilde{x}(p) \rangle$$

$$= \frac{(p+2\gamma)(p+\lambda)(p+\lambda+2\gamma)-4\sigma^2\gamma^2 (p+\lambda)}{(p+\lambda+2\gamma)(p+\lambda)+\omega^2} x(t=0) + \frac{dx}{dt}(t=0).$$

$$\tag{11.11}$$

One can check the stability of the solution of Eq.(11.1) without performing the inverse Laplace transform in (11.11). In the absence of a driving force and for zero initial conditions, $x(t=0) = 0$, the mean solution $\langle x \rangle$ should relax to zero which means that the solution of the fourth-order polynomial in p on the left hand side of (11.11) must have no roots with a positive real part. According to the Routh-Hurwitz theorem [155], this condition is obeyed for the fourth order equation $\sum_{i=0}^{i=4} a_i x^i = 0$, if the following relations hold between the coefficients a_i:

$$\text{all } a_i > 0, \quad a_1 a_4 < a_2 a_3, \quad a_0 a_3^2 < a_1 a_2 a_3 - a_1^2 a_4. \tag{11.12}$$

These stability conditions applied to Eq.(11.11) take the following form:

$$\sigma^2 < \min \left\{ 2\beta + \alpha + (1+\alpha)^2, \ (1+\alpha^{-1})(\alpha + 2\beta), \right.$$

$$\frac{1+\alpha}{2+\alpha} \left[\alpha + 2\beta + 2(1+\alpha^2) \right] \left. \right\},$$

$$(1+\alpha)^2 (\delta - \alpha)(\delta + \alpha + 2\alpha^2) < \left\{ 2(1+\alpha) \left[\delta + (1+\alpha)^2 - \sigma^2 \right] \right.$$

$$\times \left[(1+\alpha)\delta - \alpha\sigma^2 \right] - \left[(1+\alpha)\delta - \alpha\sigma^2 \right]^2 \right\} \tag{11.13}$$

where

$$\alpha = \frac{\lambda}{2\gamma}, \quad \beta = \frac{\omega^2}{4\gamma^2}, \quad \delta = \alpha + 2\beta. \tag{11.14}$$

The slightly cumbersome inequalities (11.13) define the stability conditions in the form of the relations between three parameters, σ^2, α, and β. In the case of white noise ($\sigma^2 \to \infty$, $\lambda \to \infty$ with $2D = \sigma^2/\lambda = Const$) these inequalities are reduced to the previously obtained condition $4D\gamma < 1$.

11.1.2 *Driven oscillator*

Equation (11.1) with an external periodic force $A \sin(\Omega t)$ can be rewritten as two first order differential equations

$$\frac{dx}{dt} = y, \quad \frac{dy}{dt} = -2\gamma y - 2\gamma \xi y - \omega^2 x + A \sin(\Omega t) \tag{11.15}$$

which, after averaging, take the following form

$$\frac{d}{dt} \langle x \rangle = \langle y \rangle,$$

$$\frac{d}{dt} \langle y \rangle = -2\gamma \langle y \rangle - 2\gamma \langle \xi y \rangle - \omega^2 \langle x \rangle + A \sin(\Omega t). \tag{11.16}$$

Equation (11.16) contains a new correlator $\langle \xi y \rangle$ which has to be found separately. To this end, we use the Shapiro-Loginov procedure

(2.8) which yields

$$\frac{d}{dt}\langle \xi y \rangle = \left\langle \xi \frac{dy}{dt} \right\rangle - \lambda \langle \xi y \rangle. \tag{11.17}$$

Multiplying the second of equations (11.15) by ξ, one gets after averaging

$$\left\langle \xi \frac{dy}{dt} \right\rangle = -2\gamma \langle \xi y \rangle - 2\gamma \langle \xi^2 y \rangle - \omega^2 \langle \xi x \rangle. \tag{11.18}$$

Equation (11.18) contains two new correlators, $\langle \xi x \rangle$ and $\langle \xi^2 y \rangle$. The former can be easily found using a procedure similar to (11.17)-(11.18), namely

$$\frac{d}{dt}\langle \xi x \rangle = \left\langle \xi \frac{dx}{dt} \right\rangle - \lambda \langle \xi x \rangle = \langle \xi y \rangle - \lambda \langle \xi x \rangle. \tag{11.19}$$

To find the higher-order correlator $\langle \xi^2 y \rangle$ one has to use the splitting procedure (10.12), (10.13), which gives $\langle \xi^2 y \rangle = \langle \xi^2 \rangle \langle y \rangle = \sigma^2 \langle y \rangle$. Note that this procedure becomes exact for the special case of dichotomous noise. In order to keep our calculation exact, we restrict our attention, like the authors of [151], to dichotomous noise, while for the general case of colored noise one has to use some approximations. On substituting (11.18) into (11.17) one gets

$$\frac{d}{dt}\langle \xi y \rangle = -2\gamma \langle \xi y \rangle - 2\sigma^2 \gamma \langle y \rangle - \omega^2 \langle \xi x \rangle - \lambda \langle \xi y \rangle. \tag{11.20}$$

We thus obtain a system of four equations (11.17) - (11.20) for the four variables, $\langle x \rangle$, $\langle y \rangle$, $\langle \xi x \rangle$, and $\langle \xi y \rangle$. From these equations one can easily find the fourth-order differential equation for $\langle x \rangle$

$$\frac{d^4 \langle x \rangle}{dt^4} + 2\left(\lambda + 2\gamma\right)\frac{d^3 \langle x \rangle}{dt^3} + \left[2\omega^2 + \left(\lambda + 2\gamma\right)^2 + 2\lambda\gamma - 4\gamma^2\sigma^2\right]\frac{d^2 \langle x \rangle}{dt^2}$$

$$+ 2\left[\left(\omega^2 + \gamma\lambda\right)\left(\lambda + 2\gamma\right) - 2\lambda\gamma^2\sigma^2\right]\frac{d\langle x \rangle}{dt} + \omega^2\left[\omega^2 + \lambda\left(\lambda + 2\gamma\right)\right]\langle x \rangle$$

$$= A\left[\omega^2 - \Omega^2 + \lambda\left(\lambda + 2\gamma\right)\right]\sin\left(\Omega t\right) + 2A\Omega\left(\lambda + \gamma\right)\cos\left(\Omega t\right). \tag{11.21}$$

We seek the solution of Eq. (11.21) in the form

$$\langle x \rangle = \langle x \rangle_0 + \langle x \rangle_A, \tag{11.22}$$

where the output signal $\langle x \rangle_A$ is induced by an external field, $A \sin(\Omega t)$, and $\langle x \rangle_0$ is defined by the internal dynamics. The latter was calculated in the previous section, and its Laplace transform is defined by Eq.(10.29). Let us write the solution $\langle x \rangle_A$ of the non-homogeneous equation (11.21) in the form

$$\langle x \rangle_A = a \sin(\Omega t + \phi). \tag{11.23}$$

Then, one easily finds that

$$a^2 = \frac{A^2\left(f_2^2 + \Omega^2\lambda_2^2\right)}{\left(2\Omega^2\gamma\lambda_2 - f_1 f_2 - 4\gamma^2\Omega^2\sigma^2\right)^2 + 4\Omega^2\left[\lambda_1\left(\gamma\lambda - f_1\right) - 2\lambda\gamma^2\sigma^2\right]^2} \tag{11.24}$$

and

$$\tan\phi = \frac{2\Omega\gamma f_2^2 + 4\Omega^3\gamma^2\lambda_2 - 4\Omega\gamma^2\sigma^2\left(\lambda_1\Omega^2 + \lambda\omega^2 + \lambda^2\lambda_1\right)}{f_1\left(f_2^2 + \Omega^2\lambda_2^2\right) + 4\Omega^2\gamma^2\sigma^2\left(f_1 + \lambda^2\right)} \tag{11.25}$$

where

$$\lambda_1 = \lambda + 2\gamma; \quad \lambda_2 = 2\left(\lambda + \gamma\right), \quad f_1 = \Omega^2 - \omega^2; \quad f_2 = f_1 - \lambda\lambda_1. \tag{11.26}$$

It follows from (11.24) that the amplitude of the output signal (11.23) shows a non-monotonic dependence on σ^2 and λ (stochastic resonance). The amplitude a reaches a maximum at the following value of the noise strength,

$$\left(\sigma^2\right)_{\max} = \frac{2\gamma\left(\lambda_2\Omega^2 + \lambda_1\lambda^2\right) - \left(\Omega^2 - \omega^2\right)\left(\Omega^2 - \omega^2 - \lambda\lambda_1\right)}{4\gamma^2\left(\Omega^2 + \lambda^2\right)}. \tag{11.27}$$

The dependence of the squared ratio a/A of the amplitude of the response signal to that of the external field on the correlation rate λ for $\omega = 2\gamma = 1$ and different frequencies Ω of the external field is shown in Figs. 11.1 and 11.2 for two different noise strength $\sigma^2 = 1$

Fig. 11.1 The squared ratio of the amplitude of the response signal to that of an external field as a function of the correlation rate for noise strength $\sigma^2 = 1$. The curves displayed correspond to different values of the frequency of an external field. [Reprinted figure with permission from Ref. 24. Copyright (2003) by the American Physical Society.]

and $\sigma^2 = 5$. These graphs show typical stochastic resonance, with non-monotonic behavior for the frequencies Ω close to the resonance frequency $\Omega = \omega = 1$. The maxima are more pronounced for larger noise strength. In the limit case of white Gaussian noise, $\sigma^2 \to \infty, \lambda \to \infty$ and $\sigma^2/\lambda = D$, equation (11.24) takes the following form:

$$a = A \left[\left(\Omega^2 - \omega^2 \right)^2 + 4\gamma^2 \Omega^2 \left(1 - 2D\gamma \right)^2 \right]^{-\frac{1}{2}} . \qquad (11.28)$$

The latter result can also be obtained directly from equation (11.9) with the driving force $A \sin(\Omega t)$ on the right hand side of this equation. Hence, in the presence of white noise one obtains a "dynamic" resonance slightly renormalized by white noise. The amplitude of the output signal a turns out to be a non-monotonic function of the noise strength D for white noise as well, reaching its

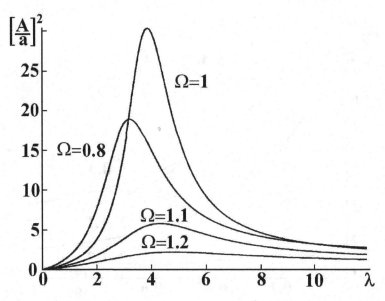

Fig. 11.2   The same as Fig. 11.1 for noise strength $\sigma^2 = 5$. [Reprinted figure with permission from Ref. 24. Copyright (2003) by the American Physical Society.]

maximum at $4D = \gamma^{-1}$. The situation becomes more complicated for colored noise where the real "stochastic" resonance occurs. For the resonant frequency $\omega = \Omega$ and $2\gamma = 1$, the amplitude of the output signal (11.24) takes the form

$$\frac{a^2}{A^2} = \frac{\lambda^2 (\lambda + 1)^2 + \Omega^2 (2\lambda + 1)^2}{\Omega^4 (2\lambda + 1 - \sigma^2)^2 + \Omega^2 \lambda^2 (\lambda + 1 - \sigma^2)^2} \qquad (11.29)$$

which is the non-monotonic function of the frequency of an external field Ω, strength σ^2 and rate λ of noise. For the special case $\sigma^2 = 1$, the amplitude increases indefinitely when $\lambda \to 0$.

11.2   Second moment for random damping

The second moments can be found in the same way as was done for the first moments. We will consider later on the stationary solutions of our equations. Therefore, in order to get non-zero stationary solutions for zero initial conditions, we take into account an additional additive noise $\eta(t)$ for the field-free case $A = 0$, while for the driven oscillator there is no need for additive noise, and we put $\eta(t) = 0$. The additive noise is taken to be white noise

$$\langle \eta(t)\, \eta(t_1) \rangle = 2\alpha\delta(t - t_1) \tag{11.30}$$

with no correlation with the multiplicative noise.

Multiplying the first of equations (11.15) by $2x$ and the second one (with additive noise) by $2y$, and averaging, one gets

$$\frac{d}{dt}\langle x^2 \rangle = 2\langle xy \rangle,$$

$$\frac{d}{dt}\langle y^2 \rangle = -4\gamma\langle y^2 \rangle - 4\gamma\langle \xi y^2 \rangle - 2\omega^2\langle xy \rangle + 2A\langle y \rangle\sin(\Omega t) + 4\alpha. \tag{11.31}$$

Analogously, multiplying equations (11.15) by y and x, respectively, summing and averaging the sum leads to

$$\frac{d}{dt}\langle xy \rangle = \langle y^2 \rangle - 2\gamma\langle xy \rangle - 2\gamma\langle \xi xy \rangle - \omega^2\langle x^2 \rangle + A\langle x \rangle\sin(\Omega t). \tag{11.32}$$

In deriving (11.31)-(11.32), we have used Eq.(2.10) for correlators containing white noise $\eta(t)$, which gives $\langle y\eta(t) \rangle = 2\alpha$ and $\langle x\eta(t) \rangle = 0$. In addition, equations (11.31)-(11.32) contain new correlators $\langle \xi y^2 \rangle$ and $\langle \xi xy \rangle$. One can calculate these and the analogous correlator $\langle \xi x^2 \rangle$ using the Shapiro-Loginov procedure (2.10) for $g = x^2$, $g = xy$ and $g = y^2$, respectively. Splitting the higher order correlators into lower order ones involves different approximations. We restrict our attention to the simplest case of white and dichotomous noises where the higher order correlators containing ξ^2 can be split by using $\xi^2 = \sigma^2$ so that, for example, $\langle \xi^2 x^2 \rangle = \sigma^2\langle x^2 \rangle$. Finally one gets

$$\frac{d}{dt}\left\langle \xi x^2 \right\rangle = 2\left\langle \xi xy \right\rangle - \lambda\left\langle \xi x^2 \right\rangle,$$

$$\frac{d}{dt}\left\langle \xi y^2 \right\rangle = -4\gamma\left\langle \xi y^2 \right\rangle - 4\gamma\sigma^2\left\langle y^2 \right\rangle - 2\omega^2\left\langle \xi xy \right\rangle$$

$$+ 2A\left\langle \xi y \right\rangle \sin\left(\Omega t\right) - \lambda\left\langle \xi y^2 \right\rangle, \tag{11.33}$$

$$\frac{d}{dt}\left\langle \xi xy \right\rangle = \left\langle \xi y^2 \right\rangle - 2\gamma\sigma^2\left\langle xy \right\rangle - 2\gamma\left\langle \xi xy \right\rangle - \omega^2\left\langle \xi x^2 \right\rangle$$

$$+ A\left\langle \xi x \right\rangle \sin\left(\Omega t\right) - \lambda\left\langle \xi xy \right\rangle.$$

We thus obtain a system of six equation (11.31)-(11.33) for six variables, $\left\langle x^2 \right\rangle$, $\left\langle y^2 \right\rangle$, $\left\langle xy \right\rangle$, $\left\langle \xi x^2 \right\rangle$, $\left\langle \xi y^2 \right\rangle$ and $\left\langle \xi xy \right\rangle$.

In the simple case of white noise one can use the simplified splitting procedure (2.10) for correlators $\left\langle \xi y^2 \right\rangle$ and $\left\langle \xi xy \right\rangle$ in Eqs. (11.31)-(11.32) which gives

$$\left\langle \xi y^2 \right\rangle \approx -8\gamma D\left\langle y^2 \right\rangle, \quad \left\langle \xi xy \right\rangle \approx -4\gamma D\left\langle xy \right\rangle. \tag{11.34}$$

Substituting (11.34) into (11.31)-(11.32), one obtains

$$\frac{d}{dt}\left\langle x^2 \right\rangle = 2\left\langle xy \right\rangle,$$

$$\frac{d}{dt}\left\langle y^2 \right\rangle = -4\gamma\left\langle y^2 \right\rangle + 32\gamma^2 D\left\langle y^2 \right\rangle - 2\omega^2\left\langle xy \right\rangle + 2A\left\langle y \right\rangle \sin\left(\Omega t\right) + 4\alpha,$$

$$\frac{d}{dt}\left\langle xy \right\rangle = \left\langle y^2 \right\rangle - 2\gamma\left\langle xy \right\rangle + 8\gamma^2 D\left\langle xy \right\rangle - \omega^2\left\langle x^2 \right\rangle + A\left\langle x \right\rangle \sin\left(\Omega t\right).$$

$$\tag{11.35}$$

Note that for white noise one can find exact expressions for higher moments of the coordinate and velocity of the form $\left\langle x^{m-n}y^n \right\rangle$. The derivative of the latter expression has the form

$$\frac{d}{dt}\left\langle x^{m-n}y^n \right\rangle = (m-n)\left\langle x^{m-n-1}y^n\frac{dx}{dt} \right\rangle + n\left\langle x^{m-n}y^{n-1}\frac{dy}{dt} \right\rangle.$$

$$\tag{11.36}$$

On substituting (11.15) (with additive noise) and (11.34) into (11.36), one obtains

$$\frac{d}{dt} \langle x^{m-n} y^n \rangle = (m-n) \langle x^{m-n-1} y^{n+1} \rangle - 2n\gamma (1 - 4\gamma D n) \langle x^{m-n} y^n \rangle$$

$$- n\omega^2 \langle x^{m-n+1} y^{n-1} \rangle + nA \sin (\Omega t) \langle x^{m-n} y^{n-1} \rangle$$

$$+ 2n(n-1)\alpha \langle x^{m-n} y^{n-2} \rangle . \tag{11.37}$$

Equation (11.37) reduces to (11.35) for the special cases $n = 0, m = 2$, $n = 2$, $m = 2$ and $n = 1$, $m = 2$, as it should be.

The correlation functions can be found along the same lines as was done in (11.31)-(11.35) for the second moments by multiplying the first of equations (11.15) by $x(t_1)$, the second one by $y(t_1)$, and averaging the resulting equations, which gives

$$\frac{d}{dt} \langle x(t_1) x(t) \rangle = \langle x(t_1) y(t) \rangle ,$$

$$\frac{d}{dt} \langle x(t_1) y(t) \rangle = -2\gamma \langle x(t_1) y(t) \rangle - 2\gamma \langle \xi(t) x(t_1) y(t) \rangle$$

$$- \omega^2 \langle x(t_1) x(t) \rangle + A \langle x(t_1) \rangle \sin (\Omega t) . \tag{11.38}$$

The new correlators $\langle \xi(t) x(t_1) y(t) \rangle$ and $\langle \xi(t) x(t_1) x(t) \rangle$ can be found by using equation (2.8) leading to

$$\frac{d}{dt} \langle \xi(t) x(t_1) x(t) \rangle = \langle \xi(t) x(t_1) y(t) \rangle - \lambda \langle \xi(t) x(t_1) x(t) \rangle ,$$

$$\frac{d}{dt} \langle \xi(t) x(t_1) y(t) \rangle = -2\gamma \langle \xi(t) x(t_1) y(t) \rangle - 2\gamma \sigma^2 \langle x(t_1) y(t) \rangle$$

$$- \omega^2 \langle \xi(t) x(t_1) x(t) \rangle - \lambda \langle \xi(t) x(t_1) y(t) \rangle$$

$$+ A \langle \xi(t) x(t_1) \rangle \sin (\Omega t) . \tag{11.39}$$

In the following sections, the formulas obtained above will be applied to different special cases.

11.3   Force-free oscillator

11.3.1   *White noise*

The stationary ($\frac{d}{dt}... = 0$) second moments of the force-free ($A = 0$) oscillator are obtained from Eq. (11.35) which gives

$$\langle xy \rangle_{eq} = 0, \quad \langle y^2 \rangle_{eq} = \omega^2 \langle x^2 \rangle, \quad \langle x^2 \rangle = \frac{\alpha}{\omega^2 \gamma (1 - 4\gamma D)}. \quad (11.40)$$

For vanishing multiplicative noise, $D = 0$, Eqs. (11.40) reduce to their standard form describing Brownian motion. In the presence of this noise, Eqs. (11.40) show that an (energetic) instability occurs when $4\gamma D > 1$. This result is different from that obtained for the case of the random frequency, where, according to (10.26), the instability condition of the second moment has the form $\omega^2 D > 2\gamma$.

11.3.2   *Dichotomous noise*

For dichotomous noise it is necessary to use the six equation (11.31)-(11.33), which we will consider in the equilibrium state ($\frac{d}{dt}... = 0$) and in the absence of an external force ($A = 0$). Their solution is

$$\langle x^2 \rangle_{eq}$$

$$= \frac{4\alpha\{(\lambda + 2\gamma) \left[4\omega^2 + \lambda (\lambda + 4\gamma)\right] + 8\lambda\gamma^2\sigma^2\}}{4\gamma (\lambda + 2\gamma) \omega^2 \left[4\omega^2 + \lambda (\lambda + 4\gamma)\right] - 8\omega^2\gamma^2\sigma^2 \left[2\omega^2 + \lambda (\lambda + 4\gamma)\right]}.$$

$$(11.41)$$

In the limit case of white noise ($\sigma^2 \to \infty$, $\lambda \to \infty$ with $\frac{\sigma^2}{\lambda} = 2D$) equation (11.41) reduces to (11.40) as it should do. For dichotomous noise, the second moment (11.41) is a non-monotonic function of both σ^2 and rate λ.

11.4   Driven oscillator

All of the preceding was related to the field-free case which corresponds to equations (11.33) with $A = 0$. From here on, we are interested in the response of the underdamped harmonic oscillator to a periodic field, and, for simplicity, we present only the case of white noise. The more cumbersome expression for dichotomous noise [24] is not given here.

11.5 Second moments

From equations (11.35) with $\alpha = 0$, one easily finds the third-order differential equation for $\langle x^2 \rangle$,

$$\frac{d^3 \langle x^2 \rangle}{dt^3} + 2\gamma \left(3 - 10\gamma D\right) \frac{d^2 \langle x^2 \rangle}{dt^2}$$

$$+ 2 \left[2\omega^2 + 4\gamma^2 \left(1 - 2\gamma D\right) \left(1 - 4\gamma D\right)\right] \frac{d \langle x^2 \rangle}{dt}$$

$$+ 8\gamma \left(1 - 4\gamma D\right) \omega^2 \langle x^2 \rangle$$

$$= 4A \left[\langle y \rangle + 2\gamma \left(1 - 4\gamma D\right) \langle x \rangle\right] \sin \left(\Omega t\right)$$

$$+ 2A \frac{d}{dt} \left[\langle y \rangle \sin \left(\Omega t\right)\right]. \tag{11.42}$$

The first moments, $\langle x \rangle$ and $\langle y \rangle = \frac{d\langle x \rangle}{dt}$, that enter Eq.(11.42) consist of two parts, $\langle x \rangle_0$ and $\langle x \rangle_a$, where the first one is the solution of the field-free equation (11.9), and the second one is defined by (11.23). In the stable region ($4\gamma D < 1$), the solution of Eq. (11.9) vanishes as $t \to \infty$, and the stationary solution of Eq. (11.42) ($\frac{d}{dt} \ldots = 0$) is obtained on substitution $\langle x \rangle = \langle x \rangle_a = a \sin \left(\Omega t + \phi\right)$ and $\langle y \rangle = \frac{d\langle x \rangle}{dt} = a\Omega \cos \left(\Omega t + \phi\right)$. Thus, one gets for the stationary value of the second moment,

$$\langle x^2 \rangle_{st} = \frac{aA}{2\omega^2} \left(\cos \phi - \frac{\Omega \sin \phi}{2\gamma \left(1 - 4\gamma D\right)}\right) \tag{11.43}$$

where a and $\tan \phi$ are defined in (11.24)-(11.25). It follows from (11.43) that the second moment $\langle x^2 \rangle_{st}$, is a non-monotonic function of the noise strength D. It turns out that for dichotomous noise the second moment shows a non-monotonic dependence of both σ^2 and λ.

11.6   Correlation functions

From equations (11.38)-(11.39) one can find the fourth order differential equation for $z \equiv \langle x(t_1) x(t) \rangle$,

$$
\frac{d^4 z}{dt^4} + 2\left(\lambda + 2\gamma\right) \frac{d^3 z}{dt^3} + \left[2\omega^2 + 4\gamma^2 + 6\lambda\gamma + \lambda^2 - 4\gamma^2 \langle \xi^2 \rangle\right] \frac{d^2 z}{dt^2}
$$

$$
+ \left[\left(2\omega^2 + 2\lambda\gamma\right)\left(\lambda + 2\gamma\right) - 4\lambda\gamma^2 \langle \xi^2 \rangle\right] \frac{dz}{dt}
$$

$$
+ \omega^2 \left[\omega^2 + \lambda\left(\lambda + 2\gamma\right)\right] z
$$

$$
= -A \left[\frac{d^2}{dt^2} + 2\left(\lambda + \gamma\right) \frac{d}{dt} + \left(\omega^2 + \lambda^2 + 2\lambda\gamma\right)\right] \langle x(t_1) \rangle \sin\left(\Omega t\right)
$$

$$
+ 2A\gamma \left(\frac{d}{dt} + \lambda\right) \langle \xi(t) x(t_1) \rangle \sin\left(\Omega t\right). \tag{11.44}
$$

In order to avoid the need for the rather long formulas, we restrict our attention to the case of white noise. With equations (2.9)-(2.10) one gets from (11.39)

$$
\langle \xi(t) x(t_1) x(t) \rangle = 0; \quad \langle \xi(t) x(t_1) y(t) \rangle = -4\gamma D \langle x(t_1) y(t) \rangle, \tag{11.45}
$$

and, the substitution of these formulas into (11.38) leads to

$$
\left[\frac{d^2}{dt^2} + 2\gamma\left(1 - 4\gamma D\right) \frac{d}{dt} + \omega^2\right] \langle x(t_1) x(t) \rangle = Aa \sin\left(\Omega t_1 + \phi\right) \sin\left(\Omega t\right). \tag{11.46}
$$

The form of the solution of equation (11.46) depends on the type of the initial conditions, which can be either $\langle x(t_1) x(t) \rangle$ at $t = t_1$ or at $t = 0$. For the latter case one assumes that the initial condition $x(t = 0)$ is not random. Then, $\langle x(t_1) x(t = 0) \rangle = x(t = 0) \langle x(t_1) \rangle$ and $\frac{d}{dt} \langle x(t_1) x(t = 0) \rangle = \frac{dx}{dt}(t = 0) \langle x(t_1) \rangle$. Solving the non-homogeneous equation (11.46) through the use of the

Green function, one obtains

$$\langle x(t_1) x(t) \rangle$$

$$= \exp\left[-\gamma(1 - 4\gamma D)t\right] \{x(t = 0) \langle x(t_1) \rangle [\cos(\omega_2 t)$$

$$+ \frac{\gamma(1 - 4\gamma D)}{\omega_2} \sin(\omega_2 t) + \frac{1}{\omega_2} \frac{dx}{dt}(t = 0) \langle x(t_1) \rangle \sin(\omega_2 t)]\}$$

$$+ Aa \sin(\Omega t_1 + \phi)$$

$$\times \int_0^t \exp\left[-\gamma(1 - 4\gamma D)(t - \theta)\right] \cos\left[\omega_2(t - \theta)\right] \sin(\Omega \theta) d\theta$$

$$\tag{11.47}$$

where $\omega_2 = \sqrt{\omega^2 - \gamma^2(1 - 4\gamma D)^2}$, and the last integral in (11.47) can be easily expressed in terms of elementary functions.

For white noise, the correlation function (11.47), like the second moment $\langle x^2 \rangle$, is a non-monotonic function of the noise strength D. For dichotomous noise the correlation function shows a non-monotonic dependence on both σ^2 and λ.

11.7 Periodically varying damping

It is instructive to compare the influence of multiplicative noise with the effects found in an oscillator with a periodically changed damping parameter described by the following equation

$$\frac{d^2 x}{dt^2} + \gamma[1 + b \sin(\Omega_1 t)]\frac{dx}{dt} + \omega^2 x = 0. \tag{11.48}$$

As already mentioned in section 7.3, equation (11.48) with periodic coefficients has the Floquet solution in the form [100]

$$x(t) = \exp(\alpha t)\psi(t)$$

$$= \exp(\alpha t) \sum_{n=0}^{\infty} \left[A_n \sin\left(\frac{n\Omega_1 t}{2}\right) + B_n \cos\left(\frac{n\Omega_1 t}{2}\right)\right] \tag{11.49}$$

where the periodic function $\psi(t)$ is expanded in a Fourier series. As is evident from (11.49), $x(t)$ vanishes at $t \to \infty$ for $\alpha < 0$, diverges for $\alpha > 0$, and remains a bounded periodic function for $\alpha = 0$. Hence, $\alpha = 0$ defines the stability boundary of the stationary solutions of equation (11.48). On the substitution of (11.49) with $\alpha = 0$ into (11.48), and comparing the harmonics in front of the sine and cosine terms, one obtains the infinite systems of linear equations for A_n and B_n which have non-zero solutions if the infinite determinant of these equations $\Delta(\alpha = 0)$ vanishes, $\Delta(\alpha = 0) = 0$. One has to truncate this determinant at some n, and afterwards to improve the result by taking into account larger values of n. Leaving only terms with $n = 1$, one obtains the following equations

$$\left(\omega^2 - \frac{\Omega_1^2}{4} + \frac{\Omega_1\gamma b}{4}\right) A_1 + \frac{\Omega_1\gamma}{2} B_1 = 0,$$

$$-\frac{\Omega_1\gamma}{2} A_1 + \left(\omega^2 - \frac{\Omega_1^2}{4} - \frac{\Omega_1\gamma b}{4}\right) B_1 = 0.$$

(11.50)

Equations (11.50) have nontrivial solutions if the determinant of these equations vanishes, which gives

$$b = \sqrt{4 + \frac{\left(4\omega^2 - \Omega_1^2\right)^2}{\gamma^2\Omega_1^2}}.$$

(11.51)

The stability boundary (11.51) of the solution $x = 0$ has a V form at $b - \Omega$ plane with a stable state located inside this curve. Equation (11.51) defines a necessary condition for the non-zero periodic solution of the periodically varying velocity in the same way as the condition $4\gamma D = 1$ defines it for a random velocity. The difference is that in the former case an external field defines the basic frequency Ω_1 of oscillation, while in the latter case the oscillations occur at the oscillator frequency ω.

Chapter 12

Nonlinear oscillator with multiplicative noise

12.1 Double-well potential (noise induced reentrant transition)

In chapter 7 we considered an overdamped noisy oscillator in a double-well potential. In this chapter the same problem is discussed for an underdamped oscillator. In the presence of damping, the equation of motion has the form

$$\frac{d^2x}{dt^2} + \gamma \frac{dx}{dt} - ax + bx^3 + \sqrt{2D}\xi(t)\,x = 0. \tag{12.1}$$

It turns out [156], that for a noisy system there is some region of positive a where reentrant transitions occur as a function of the noise strength, showing noisy oscillations for both weak and strong noise.

On passing to dimensionless variables $\tau = \gamma t$ and $y = \sqrt{b}x/\gamma$, one gets,

$$\frac{d^2y}{d\tau^2} + \frac{dy}{d\tau} - \alpha y + y^3 + \sqrt{2\Delta}\xi(t)\,y = 0 \tag{12.2}$$

with dimensionless coefficients

$$\alpha = \frac{a}{\gamma^2}, \quad \Delta = \frac{D}{\gamma^4} \tag{12.3}$$

Stability at the origin, $x = 0$, is defined by the linearized version of (12.2)

$$\frac{d^2y}{d\tau^2} + \frac{dy}{d\tau} - \alpha y + \sqrt{2\Delta}\xi\left(\gamma^{-1}\tau\right)y = 0. \tag{12.4}$$

The deviations from the trivial solution $y(\tau) = \frac{dy}{d\tau}(\tau) = 0$, which are described by equation (12.4), will lead to a new nontrivial solution when the Lyapunov index of this equation will change the sign. The Lyapunov index Λ is defined as the exponential divergence rate of neighboring trajectories [157], i.e., as

$$\Lambda = \lim_{t \to \infty} \lim_{\epsilon \to 0} \frac{1}{t} \int_0^t d\tau \frac{1}{\epsilon} \ln \frac{y(\tau + \epsilon)}{y(\tau)}. \qquad (12.5)$$

It is convenient, [158] to take the limit $\epsilon \to 0$ first. Then, after substituting in (12.5) the expansion $y(\tau + \epsilon) = y(\tau) + \epsilon \frac{dy}{d\tau} + ...$, one gets

$$\Lambda = \lim_{t \to \infty} \frac{1}{t} \int_0^t d\tau \; z(\tau) = \langle z \rangle \equiv \int z P_{as}(z) \, dz \qquad (12.6)$$

where the new variable $z = \frac{dy/d\tau}{y}$ has been introduced. In order to find the asymptotic $(\tau \to \infty)$ probability distribution function $P(z)$ for the variable z, one can, using (12.4), find the Langevin equation for z,

$$\frac{dz}{d\tau} = \alpha - z - z^2 - \sqrt{2\Delta}\xi(\tau). \qquad (12.7)$$

So far we did not make any assumption about the type of noise in the original equation (12.1). However, for different types of noise one gets different forms of the Fokker-Planck equation corresponding to the Langevin equation (12.7).

For a Gaussian white noise $\xi(x)$,

$$\langle \xi(t) \rangle = 0, \quad \langle \xi(x)\xi(x_1) \rangle = 2\delta(x - x_1), \qquad (12.8)$$

the Stratonovich form of the Fokker-Planck equation for $P(z)$ which corresponds to the Langevin equation (12.7) with noise of the form (12.8) has the following form [28]

$$\frac{\partial P}{\partial \tau} = -\frac{\partial}{\partial z}\left[(\alpha - z - z^2)P\right] + \Delta\frac{\partial^2 P}{\partial z^2} = 0. \qquad (12.9)$$

The integrable asymptotic $(\tau \to \infty)$ solution of this equation is found by the method of variation of constants, which gives

$$P_{as}(z) = N \int_{-\infty}^{z} \exp\left\{\frac{1}{\Delta}\left[\Psi(z) - \Psi(y)\right]\right\} \tag{12.10}$$

where $\Psi(y) = \alpha y - y^2/2 - y^3/3$.

On substituting (12.10) into (12.6), one obtains [156],

$$\Lambda = \frac{1}{2}\left\{\frac{\int_0^{\infty} du\sqrt{u}\exp\left[\left(\alpha+\frac{1}{4}\right)\frac{u}{\Delta} - \frac{u^4}{12\Delta}\right]}{\int_0^{\infty}\frac{du}{\sqrt{u}}\exp\left[\left(\alpha+\frac{1}{4}\right)\frac{u}{\Delta} - \frac{u^4}{12\Delta}\right]} - 1\right\}. \tag{12.11}$$

For different values of the parameters α and Δ defined in (12.3), the Lyapunov index Λ can be either positive, describing thereby the stability of the origin, $x = 0$, or negative which represents an instability.

Analysis of the function $\Lambda(\alpha, \Delta)$ shown in Fig. 12.1, shows that [156]

a) For $\alpha = -0.2$, λ changes sign once (noise-induced instability).

b) For $\alpha = 0.2$, λ changes sign twice (noise induced reentrant transition)

c) For $\alpha = 0.5$, λ is positive for all Δ (no transition).

Hence, noise can suppress oscillations by stabilizing a fixed point which was unstable in the absence of noise. The stability analysis of the linearized form (12.4) of the original equation (12.1) has to be performed by the use of the Lyapunov exponent which is the proper indicator of the transition in the nonlinear system. Such an analysis cannot be performed by studies of the finite-order moments of the linearized system. Indeed, the second-order moments of equation (12.4) with white noise are always unstable for positive α which contradicts the results obtained above.

Note that, in addition to the above considered white noise, the asymptotic distribution function $P_{as}(z)$ corresponding to the Langevin equation (12.7) was found in section 6.1.3 and 6.1.4 for the more general case of dichotomous and Poisson noises. For dichotomous noise (2.7) with two values $\xi(t) = \pm\sigma$ and the flipping

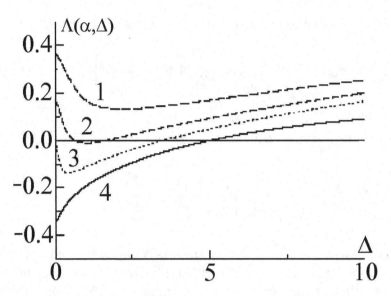

Fig. 12.1   Lyapunov exponent of a linear damped oscillator with parametric white noise as a function of the noise amplitude for different values of the control parameter α (1: $\alpha = 0.5$, 2: $\alpha = 0.2$, 3: $\alpha = 0.0$, 4: $\alpha = -0.2$). [Reprinted from Ref. 157 with permission from EDP Sciences.]

rate λ, $P_{as}(z)$ is given as

$$P_{as} = N\,(z - z_1)^{-1-\frac{\lambda}{1-\sigma^2}}$$

$$\times\,|z - z_1 + \alpha - \sigma|^{-1-\frac{\lambda}{\sigma^2+1}}\,|z - z_1 + \alpha + \sigma|^{-1-\frac{\lambda}{\sigma^2-1}} \quad (12.12)$$

where

$$z_1 = \frac{1 - \sqrt{1 + 4\alpha^2}}{2}. \quad (12.13)$$

For Poisson noise (2.11)-(2.12), $P_{as}(z)$ has the following form

$$P_{as} = N\,(z - z_1 + z_2)\,|z - z_2|^{-1-\frac{1}{wz_2}} \quad (12.14)$$

where

$$z_2 = \lambda w + 1. \quad (12.15)$$

The Lyapunov index Λ is obtained after substituting (12.12) and (12.14) into (12.6), and carrying out the numerical integration.

12.2 Duffing oscillator

The general form of power law nonlinearity in the underdamped oscillator equation of the form

$$\frac{d^2x}{dt^2} + \left[\omega^2 + \xi\left(t\right)\right] x + bx^{2n-1} = 0 \qquad (12.16)$$

has been analyzed [159]. The linear equation which corresponds to $n = 1$ was considered in chapter 10. Although some results can be obtained for the general case [159], we restrict ourselves to the special case of equation (12.16) with $n = 2$ which is called the Duffing oscillator, described by the following equation

$$\frac{d^2x}{dt^2} + \gamma\frac{dx}{dt} + \left[\omega^2 + \xi\left(t\right)\right] x + bx^3 = 0. \qquad (12.17)$$

Note that the equation of motion (12.17) of the Duffing oscillator is distinguished from the equation (12.1) for the motion in a double-well potential (which is sometimes called an "inverted Duffing oscillator") by the sign in front of the linear term, ω^2x. This leads to an important difference between these two equations, namely that the deterministic and undamped Duffing oscillator ($\gamma = \xi = 0$ in (12.17)) has only one stable point, $x = 0$. The latter equation allows an exact solution since, according to (12.17), the energy

$$E = \frac{1}{2}\left(\frac{dx}{dt}\right)^2 + \frac{\omega^2x^2}{2} + \frac{bx^4}{4} \qquad (12.18)$$

is conserved. A solution of (12.17) in terms of Jacobi elliptic functions is obtained in the energy-angle variables. Turning now to the general equation (12.17), one gets [156] the following result. Starting from a small initial condition (small energy), the amplitude of the oscillator grows exponentially with time as long as $E << \omega^4$ or $x << \omega$. For $E >> \omega^4$, the nonlinear term becomes important, and for white noise the amplitude grows as the square-root of time. The crossover from exponential to algebraic growth appears at $E \sim \omega^4$ or $x \sim \omega$, where the linear and nonlinear terms are of the same order. It is remarkable that the scaling indices defining the asymptotic time dependences $x \approx t^\alpha$ and $E \approx t^\beta$ are different for white and colored noise $\xi(t)$ in equation (12.17) in such a way that for white noise $\alpha = 1/2$ and $\beta = 2$ while for colored noise these indices are two times smaller, $\alpha = 1/4$ and $\beta = 1$.

The analysis of stability for equation (12.17) can be performed analogously to that of (12.1)-(12.11) in the previous chapter. Although in this case the origin remains the fixed point for all parameters α and Δ defined in (12.3), the Lyapunov index Λ which is equal to [156]

$$\Lambda = \frac{1}{2}\left\{ \frac{\int_0^\infty du\sqrt{u}\exp\left[\left(1-\frac{\alpha^2}{4}\right)\frac{2u}{\Delta}-\frac{u^3}{6\Delta}\right]}{\int_0^\infty \frac{du}{\sqrt{u}}\exp\left[\left(1-\frac{\alpha^2}{4}\right)\frac{2u}{\Delta}-\frac{u^3}{6\Delta}\right]} - \alpha \right\} \qquad (12.19)$$

defines the transition from the absorbing state at origin characterized by delta function form of stationary distribution to an oscillatory asymptotic state with a nontrivial form of the stationary distribution function representing a dynamic balance between energy dissipation and noise-induced energy injection. The transitions between these two states are defined by the sets of parameters α and Δ in (12.19) which make the Lyapunov index Λ equal zero.

12.3   Van der Pol oscillator

The van der Pol oscillator is described by the following equation [160]

$$\frac{d^2x}{dt^2} + \gamma\left(x^2 - 1\right)\frac{dx}{dt} + \omega_0^2 x = 0. \qquad (12.20)$$

The nonlinear damping term in (12.20) describes the energy exchange with a surrounding medium in such a way that energy is dissipated when $|x| > 1$, and fed-in when $|x| < 1$. This intrinsic periodicity is a mechanism responsible for the fact that all but one initial conditions will lead to closed stable trajectories in the phase space $x, \frac{dx}{dt}$ (limit cycles) The only other asymptotic state is a point

$x = \frac{dx}{dt} = 0$ which will be achieved only if one starts from this point. All other initial conditions will lead to limit cycles.

Consider first an additive constant force acting on van der Pol oscillator [161]

$$\frac{d^2x}{dt^2} + (x^2 - 1)\frac{dx}{dt} + x = A \qquad (12.21)$$

where by using the appropriate units for x and t, we put $\omega_0^2 = \gamma = 1$.

The bias force A in (12.21) perturbs the intrinsic periodicity of the system and introduces a new fixed point $x = A$, $\frac{dx}{dt} = 0$, which turns out to be stable for $|A| > 1$. For $|A| < 1$, an asymptotic solution, as in (12.20), has the form of a limit cycle, i.e., the bifurcation point is $x = A_c = 1$, $\frac{dx}{dt} = 0$.

Let us generalize equation (12.21) by introducing the white noise $\xi(t)$,

$$\frac{d^2x}{dt^2} + (x^2 - 1)\frac{dx}{dt} + x = A + \sqrt{2D}\xi(t). \qquad (12.22)$$

The transition from (12.21) to (12.22) means that the control parameter A is now fluctuating, and the stochastic bifurcation process between the dissipative and oscillating mechanisms is different from the deterministic one. It turns out [161] that

1. For both deterministic and noisy systems the bifurcation point A_c, which divides two types of solutions has a marginal stability, which means that it takes a relaxation time $\tau \sim |A - A_c|^{-1}$ to approach this point (critical slowing-down similar to that in equilibrium systems near the critical points).

2. The location of bifurcation points A_c slightly depends on the noise intensity D, $A_c = f(D)$ in such a way that limit cycles exist up to A larger than $A_c = 1$ for $D = 0$.

3. The latter phenomenon is connected with the noise-induced periodicity caused by white noise (without an external periodic signal!) which is different from the internal periodicity of limit cycles of deterministic systems.

Stochastic noise can be added to the deterministic system (12.20) not as additive noise, as in (12.22), but as multiplicative noise

$$\frac{d^2x}{dt^2} + \gamma \left(x^2 - 1\right) \frac{dx}{dt} + \omega_0^2 x + \sqrt{2D}\xi\left(t\right) x = 0. \qquad (12.23)$$

An interesting phenomenon has been found from the analysis of this equation [162]. Since the deterministic parameter ω_0^2 determines the period of a simple harmonic oscillator and strongly influences the period of limit cycles, its randomness leads to a new phenomenon, namely the period of oscillations is found to be a decreasing function of noise intensity. This phenomenon of noise-induced speeding-up is in contrast to previously mentioned critical slowing-down.

Let us consider now the van der Pol oscillator subjected to an external periodic field $A\sin\left(\Omega t\right)$,

$$\frac{d^2x}{dt^2} + \gamma \left(x^2 - 1\right) \frac{dx}{dt} + \omega_0^2 x = A\sin\left(\Omega t\right). \qquad (12.24)$$

The solution of equation (12.24) is, in general, a sum of two periodic solutions corresponding to the natural frequency of the oscillator ω_0 and the frequency of the driving force Ω. When Ω approaches ω_0, the phenomenon of frequency locking occurs, namely, the solution of (12.24) with the frequency ω_0 will disappear at some finite value of $|\Omega - \omega_0|$ (and not at $\Omega = \omega_0$), and the remaining solution will be that with the frequency Ω. Performing the first approximation of the Krylov-Bogolyubov averaging procedure [163], one can show [160] that for $\frac{A^2}{\gamma^2\Omega^2} > \frac{32}{27}$, the frequency locking solution is stable for the following relation between the amplitude A and frequency Ω of the external field,

$$\mid \Omega - \omega_0 \mid < \frac{\gamma}{4}\sqrt{\left(\frac{2A^2}{\gamma^2\Omega^2} - 1\right)}. \qquad (12.25)$$

For $\frac{A^2}{\gamma^2\Omega^2} < 1$, the frequency locking solution is stable for all values of Ω up to a threshold value of Ω_c found in [160]. An approximate analysis of the frequency locked solution has also been performed in the presence of additive white noise $\xi\left(t\right)$ to (12.24)[160].

Chapter 13

In the future ...

One hundred years have passed since the explanation of the phenomenon of Brownian motion by Einstein, Smoluchowski and Langevin. Like other great inventions such as general relativity, a need was generated for a special mathematical description. It has taken the new concept of stochastic differential equations to express the molecular-kinetic aspect of this phenomenon. The fluctuation force exerted on the Brownian particle by the molecules of a surrounding medium was depicted by an additive noise in the differential equation of motion of a Brownian particle. During the last hundred years, a large body of work has been devoted to the modelling of different phenomena in physics, chemistry, biology and sociology through the use of additive and multiplicative noise in differential equations. This noise has an internal or an external origin, respectively. A dynamic oscillator is the simplest toy model for different phenomena in Nature, and taking into account the surrounding noise makes these models more adequate. Many new phenomena have been found during the past hundred years, and their number is growing like a snowball. The most impressive is the deep relationship between determinism and stochasticity - their general meaning, like that of relativity and quantum mechanics, is certainly going beyond the scope of physics. The widely studied phenomena of "deterministic chaos" [164] and "stochastic resonance" (SR) [114] might sound internally contradictory, consisting of half-deterministic and half-random terms. In fact, deterministic chaos denotes a random type of behavior in deterministic systems, while SR shows deterministic-like behavior in random systems. These peculiar features show that determinism and randomness are complementary, rather than contradictory, phenomena [165]. The peculiarity of SR lies in the fact that noise, which usually appears as a destructive factor, may play a constructive role. The reader will certainly enjoy (together with the author) the title of a recent article [166] "Noise is good for the brain". It turns out, in particular, that the human brain make use of externally-added noise to improve the detection of a weak visual signal [167], and to modulate attention switching between spatial location [168]. Stochastic phenomena in Nature still hold many surprises in store ...

Bibliography

[1] M. Moshinski and Yu F. Smirnov, *The Harmonic Oscillator in Modern Physics* [Harwood, The Hetherlands, 1996].

[2] A. Einstein in R. H. Furth ed. *Investigations on the Theory of the Brownmian Motion* [Dover, New York, 1954].

[3] M. Smoluchowski, Ann. Phys., **21**, 756 (1906).

[4] P. Langevin, Compt. Rend. **146**, 530, 1908

[5] L. D. Landau and E. M. Lifshitz, *Statistical Physics,* [Pergamon 1980].

[6] A. H. Nayfeh, *Perturbation Methods* [Wiley, New York 1974].

[7] S. C. Venkataramani, T. M. Antonsen, Jr., E. Ott, amd J. C. Sommerer, Physica D **96**, 66 (1996).

[8] R. Graham, M. Hoihenerbach, and A. Schenze, Phys. Rev. Lett., **48**, 1396 (1982).

[9] W. Kohler and G. C. Papanicolaou, in *Springer Lectures in Physics,* **70** [Springer, New York 1977]

[10] V. L. Klyatskin and V. I. Tatarskii, Soviet Physics - Usp. **16**, 494 (1974).

[11] M. Turelli, *Theoretical Population Biology* [Academic, New York, 1977].

[12] H. Takayasu, A-H. Sato, and M. Takayasu, Phys. Rev. Lett., **79**, 966 (1997).

[13] V. I. Klyatskin, *Stochastic Equations and Waves in Randomly Inhomogeneous Media,* [Nauka,Moscow, 1980] (In Russian); V. I. Klyatskin, Soviet Physics - Acoustic **26**, 113 (1980).

[14] M. Gitterman, Phys. Rev. E **67,** 057103 (2003).

[15] M. Gitterman, Physica A **352**, 305 (2005).

[16] B. West and V. Seshadri, J. Geophys. Res. **86**, 4293 (1981).

[17] M. Gitterman, Phys. Rev. E 70, 036116 (2004).

[18] A. Onuki, J. Phys : Condens. Matter **9**, 6119 (1997).

[19] J. M. Chomaz and A. Couairon, Phys. Fluids, **11**, 2977 (1999).

[20] T. E. Faber, *Fluid Dynamics for Physicists* [Cambridge University Press, Cambridge, 1995].

[21] F. Heslot and A. Libchaber, Phys. Scr. **T9**, 126 (1985).

[22] A. Saul and K. Showalter, *Ocsillations and Travel Waves in Chemical Systemns*, ed. R.J. Field and M. Burger [Wiley, New York, 1985].

[23] M. Gitterman, B. Ya. Shapiro, and I. Shapiro, Phys. Rev. E **65**, 174510 (2002).

[24] M. Gitterman, Phys. Rev. E **69**, 041101 (2004).

[25] K. Furutsu, J. Res. NBS, D **67**, 303 (1963) ; E.A. Novikov, Sov. Phys.-JETP, **20**, 1290 (1965).

[26] V. E. Shapiro and V. M. Loginov, Physica A **91**, 563 (1978).

[27] E. Nelson, *Dynamical Theories of Brownian Motion* [Princeton, 1967].

[28] N. G. van Kampen, *Stochastic Processes in Physics and Chemistry* [North-Holland, Amsterdam 1992].

[29] Yu. L. Klimontovich, Physica A **163**, 515 (1992).

[30] J. J. Brey, J. M. Casado, and M. Morillo, Z. Phys. B. **66**, 263 (1987).

[31] P. Jung and P. Hanggi, Phys. Rev. A **35**, 4464 (1987).

[32] P. Hanggi and P. Jung, Adv. Chem. Phys. **89**, 239 (1995).

[33] S. Z. Ke, D. J. Wu, and L. Cao, Eur. Phys. J. B **12**, 119 (1999).

[34] J-P. Bouchaud and A. Georges, Phys. Rep., **195**, 127 (1990).

[35] R. Meltzer and J. Klafter, Phys. Rep., **333**, 1 (2000).

[36] S. I. Denisov and W. Horsthemke, Phys. Rev. E **62**, 7729 (2000).

[37] F. Lillo and R. N. Mantegna, Phys. Rev. E **61**, R4675 (2000).

[38] M. Gitterman and V. Steinberg, Phys. Fluids, 23, 2154 (1980).

[39] R. Kubo, Rep. Prog. Phys. 29, 255 (1966).

[40] H. Xia, O. Ishibara, and A. Hirose, Phys. Fluids, B **5**, 2892 (1993).

[41] M. W. Reeks, Phys. Fluids, **31**, 1314 (1988).

[42] M. Gitterman, Phys. Rev. E **52**, 303 (1995) ; Physica A **221**, 330 (1995).

[43] E. Fermi, Phys. Rev. **75**, 1169 (1949).

[44] C. Van den Broeck and R. Kawai, Phys. Rev. E **57**, 3866 (1998).

[45] V. E. Shapiro, Phys. Rev. E **48**, 109 (1993).

[46] L. Landau and E. Lifshitz, *Mechanics* [Pergamon, 1976].

[47] A. Fulinski and P.F. Gora, Phys. Rev. E **48**, 3510 (1993).

[48] F. Liu, B. Hu, and W. Wang, Phys. Rev. E **63**, 031907 (2001).

[49] C. Zhou, J. Kurtis, I. Z. Kiss, and J. L. Hudson, Phys. Rev. Lett. **89**, 014101 (2002).

[50] B. Allen and J. D. Romano, Phys. Rev. D **59**, 102001 (1999).

[51] Y. El-Mohri, L. E. Antonuk, Q. Zhao, M. Maolinbay, X. Rong, K.-W. Jee, S. Nassif, and C. Cionka, Med. Phys. **27**, 1855 (2000).

[52] K. S. Fa, Chem. Phys. **287**, 1 (2003).

[53] M. R. Young and S. Singh, Phys. Rev. A **38**, 238 (1988).

[54] M. Gitterman, J. Phys. A **32**, L293 (1999).

[55] V. Berdichevski and M. Gitterman, Phys. Rev. E **60**, 1494 (1999).

[56] R. Benzi, S. Sutera, and A. Vulpiani, J. Phys. A **14**, L453 (1981) ; C. Nicolis, Tellus **34**, 1 (1982).

[57] S. Sinha, Physica A **270**, 204 (1999).

[58] H. Gang, T. Ditzinger, C.Z. Ning, and H. Haken, Phys. Rev. Lett. **71**, 807 (1993) : J.L. Cabrera, J. Corronogoitia, and F.J. de la Rubia, Phys. Rev. Lett. **82**, 2816 (1999).

[59] J.J. Collins, C.C. Chow, A.C. Capela, and T.T.Imhoff, Phys. Rev. E **54**, 5575 (1996).

[60] C.R. Doering and J.C. Gadoua, Phys. Rev. Lett. **69**, 2318 (1992).

[61] A. Fulinski, Phys. Rev. E **52**, 4523 (1995).

[62] V. Berdichevsky and M. Gitterman, Europhys. Lett. **36**, 161 (1996).

[63] A.V. Bazikin and K. Seki, Europhys. Lett. **40**, 117 (1997).

[64] V. Bezak, Czech. J. Phys. **48**, 529 (1998).

[65] A.V. Bazikin and K. Seki, and F. Shibata, Phys. Rev. E **57**, 6555 (1998).

[66] A. Fuliinski and P. F. Gora, J. Stat. Phys. **101**, 483 (2000).

[67] Y. Jin, W. Xu, M. Xu, and T. Fang, J. Phys. A **98**, 3733 (2005).

[68] M. I. Dykman, D. G. Luchinsky, P. V. E. McClintock, N. D. Stein, and N. G. Stocks, Phys. Rev. A **46**, R1713 (1992).

[69] L. Cao and D. J. Wu, Europhys. Lett. **61**, 593 (2003).

[70] P. F. Gora, Europhys. Lett. 66, 462 (2003).

[71] B.-Q. Ai, X.-J. Wang, G.-T. Liu, and L.-G. Liu, Phys. Rev. E **67**, 022903 (2003).

[72] W. Horsthemke and M. Malek-Mansour, Z. Phys. B **24**, 307 (1976).

[73] L. Arnold and W. Horsthemke, and R. Lefewer, Z. Phys. B **29**, 867 (1978).

[74] W. Horsthemke, and R. Lefewer, *Noise Induced Phase Transitions* [Springer, Berlin 1984].

[75] A. V. Soldatov, Mod. Phys. Lett. **7**, 1253 (1993).

[76] C. Van den Broeck and P. Hanggi, Phys. Rev. A **30**, 2730 (1983).

[77] K. Kitahara,W. Horsthemke, R. Lefever, and Y. Inara, Prog. Theor. Phys. **64**, 1233 (1980); K. Kitahara,W. Horsthemke, and R. Lefever, Phys. Lett. A **70**, 377 (1979).

[78] J. M. Sancho, M. San Miguel, L. Pesquera, and M. A. Rodriquez, Physica A **142**, 532 (1987).

[79] F. J. de la Rubia, Phys. Lett. A **110,** 17 (1985).

[80] I. Dayan, M. Gitterman, and G. H. Weiss, Phys. Rev. A **46**, 757 (1992); M. Gitterman, and G. H. Weiss, Physica A **70**, 107 (1993).

[81] R. N. Mantegna and B. Spagnolo, Phys. Rev. Lett. **76**, 563 (1996).

[82] N. V. Agudov, A. A. Dubkov, and B. Spagnolo, Physica A **325**, 144 (2003).

[83] J. Mills, Phys. Lett. A **133**, 295 (2000).

[84] M. Gitterman, J. Phys A **34**, L355 (2001).

[85] V. Berdichevsky and M. Gitterman, J. Phys. A **31**, 9773 (1998).

[86] V. Berdichevsky and M. Gitterman, Phys. Rev. E **59**, R9 (1999).

[87] M. Gitterman, Phys. Rev. E **65**, 031103 (2002).

[88] R. Graham and A. Schenze, Phys. Rev. A **25**, 1731 (1982).

[89] R. Graham, Phys. Rev. A **25**, 3234 (1982).

[90] R. Graham, Phys. Rev. A **10**, 1762 (1974).

[91] H. Brand and A. Schenze, Phys. Rev. A **20**, 1628 (1979).

[92] S Kabashima and T. Kawakubo, in *Systems Far From Equilibrium,* ed. L. Garrido, Lecture Notes in Physics **132** [Spriger, Berlin 1980].

[93] D.-J. Wu, L. Cao, and S.-Z. Ke, Phys. Rev. E **50**, 2496 (1994).

[94] J. M. Sancho, M. San Miguel, S. L. Katz, and J. D. Gunton, Phys. Rev. A **26**, 1589 (1982).

[95] Ya Jia and J.-R. Li, Phys. Rev. E **53**, 5786 (1996); Physica A **252**, 417 (1998).

[96] D. Mei, C. Xie, and L. Zhang, Phys. Rev. E **68**, 051102 (2003).

[97] E. Cortes and K. Lindenberg, Physica A **123**, 99 (1984).

[98] A. Teubel, U. Behn, and A. Kuhnel, Z Phys. B **71**, 392 (1988).

[99] I. S. Gradstein and I. M. Ryzhik, *Tables of Integrals,Series and Products,* [Academic, Boston 1994].

[100] E. T. Whittaker and G. N. Watson, *A Course of Modern Analysis* [Cambridge University Press, 1027].

[101] P. Jung, Z. Phys. B **76**, 521 (1989).

[102] H. Risken, *The Fokker-Planck Equation* [Springer, Berlin 1996].

[103] F. Bass and M. Gitterman, Radiophysics Radioastronomy, **61**, 71 (2001).

[104] J. Masoliver, B. J. West, and K. Lindenberg, Phys. Rev. A **35**, 3086 (1987).

[105] Ya Jia and J.-R. Li, Phys. Rev. E **53**, 5764 (1996).

[106] A.J. R. Madureira, P. Hanggi, V. Buonomano, and W. A. Rodrigez Jr., Phys. Rev. E **51**, 3849 (1995).

[107] Y. Jin and W. Xu, Chaos, Soliton, Fractals, **23**, 275 (2005).

[108] L. Gammaitoni, F. Marchesoni, and S Santucci, Phys. Rev. Lett. **74**, 1052 (1995).

[109] T. Iwai, Physica A **300**, 350 (2001).

[110] D. Dan and A. M. Jayannavar, Physica A **345**, 404 (2005).

[111] P. Jung, Phys. Rep. **234**, 175 (1993).

[112] F. Moss, *Contemporary problems in statistical physics,* ed. G. H. Weiss [SIAM Philadelphia 1994]. p.205

[113] B. McNamara and K. Wiesenfeld, Phys. Rev. A **39**, 4854 (1988).

[114] L. Gammaitoni, P. Hanggi, P. Jung, and F. Marchesoni, Rev. Mod. Phys. **70**, 233 (1998).

[115] Y. Jia, S.-N. Yu, and J-R. Li, Phys. Rev. E **62**, 1869 (2000).

[116] Y. Jia, X.-p. Zheng, X.-m. Hu, and J-r. Li, Phys. Rev. E **63**, 031107 (2001).

[117] X. Luo and S. Zhu, Phys. Rev. E **67**, 021104 (2003).

[118] R. Rozenfeld, A. Neiman, and L. Schimansky-Geier, Phys. Rev. E **62**, 051107 (2001).

[119] J.-H. Li, Phys. Rev. E **66**, 031104 (2002).

[120] E. Guardia and M. San Miguel, Phys. Lett. A **109**, 9 (1985).

[121] P. Pechukas and P. Hanggi, Phys. Rev. Lett. **73**, 2772 (1994).

[122] J. H. Li, D. Y. Xing, J. M. Dong, and B. Hu, Phys. Rev. E **60**, 1324 (1999).

[123] J. H. Li, B. Hu, D. Y. Xing, and J. M. Dong, Phys. Rev. E **60**, 6443 (1999).

[124] C. Van den Broeck, Phys. Rev. E **47**, 4579 (1993).

[125] J. C. Fletcher, S. Havlin, and G. H. Weiss, J. Stat. Phys. **51**, 215 (1988).

[126] M. Gitterman and G. H. Weiss, J. Stat. Phys. **74**, 941 (1988).

[127] J. J. Brey and J. Casado- Pascual, Physica A **212**, 123 (1994).

[128] J. M. Porra, Phys. Rev. E **55**, 6533 (1997).

[129] M. I. Dykman, D. G. Luchinsky, R. Mannela, P. V. E. Nc Clintock, N. D. Stein, and N. G. Stocks, Il Nuovo Cimento D **17**, 661 (1995).

[130] L. H'walisz, P. Jung, P. Hanggi, P. Talkner, and L. Shimansky-Geier, Z. Phys. B **77**, 471 (1989).

[131] K.-G. Wang and J. Masoliver, Physica A **231**, 615 (1996), I. Masoliver and J. M. Porra, Phys. Rev. E **48**, 4309 (1993).

[132] M. C. Wang and G. E. Uhlenbeck, Rev. Mod. Phys. **17**, 323 (1945).

[133] J. J. Stoker, Nonlinear Vibrations in Mechanical and Electrical Systems. [Interscience, New York 1950].

[134] G. Young, Proc. Edinburgh Math. Soc. **32**, 81 (1913).

[135] M. Gitterman, R. J. Shrager, and G. H. Weiss, Phys. Lett. A **142**, 841 (1989).

[136] C. Zerbe, P. Jung, and P. Hanggi, Phys. Rev. E **49**, 3626 (1999).

[137] A. R. Bulsara, K. Lindenberg, and K. E. Shuler, J. Stat. Phys. **27**, 787 (1982).

[138] A. B. Budgor, K. Lindenberg, and K. E. Shuler, J. Stat. Phys. **15**, 375, (1976).

[139] M. I. Dykman and M. A. Krivoglaz, Physica A **104**, 495 (1980) ; M. I. Dykman, S. M. Soskin, and M. A. Krivoglaz, Physica A **133**, 53 (1985) ; M. I. Dykman, et. al. Phys. Rev. A **37**, 1303 (1988).

[140] P. S. Landa and P. V. E. McClintock, J. Phys. A **33**, L433 (2000).

[141] P. Jung, Pys. Rep. 234, 175 (1993).

[142] I. I. Blekhman and P. S. Landa, Int. J. Nonlin. Mech. **39**, 421 (2004).

[143] V. N. Chizhevsky, E. Smeu, and G. Giacomelli, Phys. Rev. Lett. **91**, 220602 (2003).

[144] A. A. Zaikin, L. Lopez, J. P. Baltanas, J. Kurth, and M. A. F. Sanjuan, Phys. Rev. E **66**, 011106 (2002).

[145] K. Lindenberg, V. Seshadri, and B. West, Physica A **105**, 445 (1988)}.

[146] S. K. Banik and D. S. Ray, J. Phys. A **31**, 3937 (1998).

[147] B. V. Bobryk and A. Chreszczyk, Physica A 315, 225 (2002).

[148] W. Chang, J. Casademunt, and J. Vinals, Phys. Fluids, A **5**, 3147 (1993).

[149] R. Berthet, A. Petrossian, S. Residori, B. Roman, and S. Fauve, Physica D **174**, 84 (2003).

[150] M. F. Dimentberg, *Nonlinear Stochastic Problems of Mechanicasl Oscillations* [Nauka, Moscow 1980], in Russians.

[151] R. C. Bourret, H. Frish, and A. Pouquet, Physica **65**, 303 (1973).

[152] M. A. Leibowitz, J. Math. Phys. **4**, 852 (1963).

[153] J. Barre and T. Dauxaois, Europhys. Lett. **55**, 154 (2001).

[154] I. M. Lifshiz, S. A. Gredeskul, and L. A. Pastur, *Introduction to the Theory of Disorder Systems* [Wiley, New York 1988].

[155] S. Roul, "Stable Polynomials." §10.13 in *Programming for Mathematicians* [Springer-Verlag, Berlin 2000].

[156] K. Mallick and P. Marcq, Eur. Phys. J. B **38**, 99 (2004); ibid. **36**, 119 (2003).

[157] L. Arnold, *Random Dynamic Systems,* [Springer, Berlin, 1998].

[158] L. Tessieri and F. M. Izrailev, Phys. Rev. E **62**,3090 (2000).

[159] K. Mallick and P. Marcq, Phys. Rev. E **66**, 041113 (2002).

[160] P. Hanggi and P. Riseborough, Am. J. Phys. **51**, 347 (1982).

[161] H. K. Leung, Physica A **254**, 146 (1998).

[162] H. K. Leung, Physica A **221**, 340 (1995).

[163] N. Krylov and N. Bogoliubov, Annals Math. Studies , No **11** [Princeton, 1943].

[164] P. Berge, Y. Pomeau, and C. Vidal, *Order within Chaos* [Wiley, New York 1984].

[165] M. Gitterman, Eur. J. Phys. **23**, 119 (2001).

[166] F. Moss, Physics World, February 1977, p. 15.

[167] K. Kitajo, D. Nozaki, L. W. Ward, and Y. Yamamoto, Phys. Rev. Lett. **90**, 218103 (2003).

[168] K. Kitajo, K. Yamanaka, L. W. Ward, and Y. Yamamoto, in Fluctuations and Noise, Proc. of SPIE, vol. 5841, ed. N. G. Stokes,et. al. [SPIE, Bellingham 2005], p. 49.

Index